MW00335637

Advance Praise for *Innovating for Diversity*

"DEI is top of mind for virtually all executives regardless of the sector that they are leading in. This book highlights the key principles that are critical to success in any change effort: courage, leadership, collaboration, and trust. With compelling and detailed case studies, readers will learn how to put these principles into action."

—Jane Wei-Skillern
Senior Fellow, Center for Social Sector Leadership
UC Berkeley Haas School of Business

"The ultimate reward for truly innovating to form more diverse teams is a culture of belonging, which is absolutely necessary for any company that wants dazzling performance. I was excited and inspired by the case studies that show how great leaders are shifting from just a compliance view of DE&I to building teams where everyone is motivated to be exponential."

—Earl Newsome
Vice President & Chief Information Officer, Cummins
Founder, TechPACT

"This book dispels the notion that innovation and diversity are incompatible. It shatters the long-held belief that change must be slow. It demonstrates in tangible ways using present-day examples how companies large and small can radically improve their diversity efforts and in so doing enhance the bottom line."

—Larry Quinlan
Board Director and Former Global Chief Information Officer
Deloitte

"When it comes to DEI, saying the right thing is easier than doing the right thing. Through eye-opening stories and data, this book specifies the mindsets, metrics, and methods to realize growth pathways for people and businesses of all sizes. The authors provide perceptive questions to discern root causes of deleterious fixed practices and principles. Equally important, they illustrate the leadership behaviors plus organizational tools, which create conditions for the maximum number of diverse people to flourish and collectively accelerate innovation."

—Roselinde Torres
Leadership Expert and TED Speaker

"The authors of Innovating for Diversity bring us inspiring examples of corporate leaders courageous enough to question established practices and make substantive changes in the way they recruit, hire, develop, and advance employees. In doing so, these innovators teach us a valuable lesson: Creating truly diverse, inclusive, and innovative workplaces is good for people—and good for business."

—Maria Flynn
President and CEO
Jobs for the Future

"Meaningful diversity and equity practices are essential to the success of any organization, and I can think of no better guides through this complex process than Bertina and Susanne. Innovating for Diversity should be on the bookshelf of every manager and business leader in this country. The case studies are unique and illuminating, the advice is practical and adaptable, and readers will no doubt walk away inspired and committed to create work cultures grounded in inclusion."

—Dr. Tarika Barrett
CEO
Girls Who Code

"Having been on my own journey through life and career as a leader who happens to be Latino, I have my own stories to tell. Often, we get so caught up in the emotion and the cause, that data and facts are overlooked. I was moved and inspired by this book that brings the business cases, the stories, and for me the cultural root causes to life. We can keep talking DEIB, but this is a timely read of real wisdom and data that will impact the future of life and work in America."

—Guillermo Diaz Jr
Founder and CEO Conectado, Chair HITEC,
Board Member, Former CIO Cisco Systems

"I was so moved by the candid conversations of the executives featured in this book, and by how the authors captured their authentic, vulnerable emotions. The reflections offered in these pages should serve as a roadmap for companies who are striving to develop meaningful, measurable, and impactful DEI strategies."
—Viola Maxwell-Thompson
CEO and Board Director

"This timely, inspiring book holds a mirror to the business world, lifting the lid on poorly implemented DEI strategies and showing the pitfalls before guiding the way to success through examples, hard data, and fresh ideas. The result is a powerful case for the courage and humility to lean into our humanity so we can reap the full benefits of more genuinely diverse, kinder, safer, more creative, and more exciting workplaces."
—Peter Mousaferiadis
Founder and CEO
Cultural Infusion

"In Innovating for Diversity [the authors] explore the humble but crucial role that diversity plays in driving innovation and vice versa. Through a series of case studies of large enterprises and small businesses, the authors provide a roadmap for organisations looking to foster a more diverse culture while identifying and dismantling systemic barriers to innovation. With its timely and relevant insights, Innovating for Diversity is a must-read for business leaders focused on increasing retention and driving competitive advantage in today's world."
—Rajesh Jethwa
Chief Technology Officer
Digiterre

"The authors offer a unique presentation, which makes inclusion initiatives more relevant to those across the professional spectrum, not solely those in the C-Suite, based on one key tenet: that acknowledging the existence of imperfections is the first step to allowing real innovation to flourish. This book allows readers the opportunity to challenge existing presuppositions and think critically about future initiatives."
—Richard Gardner
CEO
Modulus

"As [the authors] state early in the book, 'diversity drives innovation.' Innovation is how we compete in an increasingly global economy. This is just one of many reasons that DEI has finally gotten the attention it deserves over the past few years, and this is one of the most timely and pivotal books I think has been released on the topic yet! Throughout these pages, these two women break down time and again the cutting edge of this rapidly expanding and crucial piece of business. With this book, how DEI works, and how to maximize it for your business (small or large) will make sense! It is well worth the read."

—Stephanie Scheller
Founder
Grow Disrupt

"[The authors] have successfully presented a framework for DEI and not simply a checklist of things to do. [The] book helps executive leaders to see the fundamental reasons why DEI is imperative beyond the monetary benefits. . .This book is a beautiful symbiosis of diversity in action."

—Meiko S. Patton
Author

"Innovating for Diversity blesses us with powerfully well-thought-out and practical strategies that explore the deepest issues confronting organizations across individuals of all sizes in the area of Diversity and Inclusion. More importantly it helps us put several issues that occur in our organizations into perspective. This book has fundamental and empowering information that we all need."

—Omu Obilor
CEO
Thrive with Omu

Innovating for Diversity

Innovating
for Diversity

Lessons from Top Companies
Achieving Business Success
through Inclusivity

Bertina Ceccarelli
Susanne Tedrick

WILEY

Copyright © 2023 by Bertina Ceccarelli and Susanne Tedrick LLC. All rights reserved.

Published by John Wiley & Sons, Inc., Hoboken, New Jersey.
Published simultaneously in Canada and the United Kingdom.

ISBN: 978-1-119-90989-7
ISBN: 978-1-119-90991-0 (ebk)
ISBN: 978-1-119-90990-3 (ebk)

No part of this publication may be reproduced, stored in a retrieval system, or transmitted in any form or by any means, electronic, mechanical, photocopying, recording, scanning, or otherwise, except as permitted under Section 107 or 108 of the 1976 United States Copyright Act, without either the prior written permission of the Publisher, or authorization through payment of the appropriate per-copy fee to the Copyright Clearance Center, Inc., 222 Rosewood Drive, Danvers, MA 01923, (978) 750-8400, fax (978) 750-4470, or on the web at www.copyright.com. Requests to the Publisher for permission should be addressed to the Permissions Department, John Wiley & Sons, Inc., 111 River Street, Hoboken, NJ 07030, (201) 748-6011, fax (201) 748-6008, or online at www.wiley.com/go/permission.

Trademarks: WILEY and the Wiley logo are trademarks or registered trademarks of John Wiley & Sons, Inc. and/or its affiliates, in the United States and other countries, and may not be used without written permission. All other trademarks are the property of their respective owners. John Wiley & Sons, Inc. is not associated with any product or vendor mentioned in this book.

Limit of Liability/Disclaimer of Warranty: While the publisher and authors have used their best efforts in preparing this book, they make no representations or warranties with respect to the accuracy or completeness of the contents of this book and specifically disclaim any implied warranties of merchantability or fitness for a particular purpose. No warranty may be created or extended by sales representatives or written sales materials. The advice and strategies contained herein may not be suitable for your situation. You should consult with a professional where appropriate. Further, readers should be aware that websites listed in this work may have changed or disappeared between when this work was written and when it is read. Neither the publisher nor authors shall be liable for any loss of profit or any other commercial damages, including but not limited to special, incidental, consequential, or other damages.

For general information on our other products and services or for technical support, please contact our Customer Care Department within the United States at (800) 762-2974, outside the United States at (317) 572-3993 or fax (317) 572-4002.

If you believe you've found a mistake in this book, please bring it to our attention by emailing our Reader Support team at wileysupport@wiley.com with the subject line "Possible Book Errata Submission."

Wiley also publishes its books in a variety of electronic formats. Some content that appears in print may not be available in electronic formats. For more information about Wiley products, visit our web site at www.wiley.com.

Library of Congress Control Number: 2022949493

Cover image: © lvnl/Adobe Stock Photos
Cover design: Wiley

SKY10042782_021523

To anyone who has ever had the courage to challenge the rules when they weren't fair.

About the Authors

Bertina Ceccarelli—Bertina is on a mission to advance racial and gender equity in the tech industry and disrupt the status quo to build a more inclusive workplace. As the CEO of NPower, one of the most successful nonprofits in North America committed to helping young adults and military-connected individuals launch tech careers, she to breaks down barriers to social and economic mobility. She is endlessly inspired by the life journeys of NPower alumni, and by the forward-looking corporate employers who see brilliance where others see limitations. Under her leadership NPower has grown its budget five-fold in the last six years and today serves over 2,000 individuals annually.

As a leader, she understands that any organization devoted to advancing diversity and equity must itself model an inclusive workplace, providing opportunities for growth and leadership at all levels. Bertina has been intentional about building a team of extraordinary colleagues who bring their deep professional expertise as well as their personal experiences to the mission. NPower's team demonstrates how the power of diversity delivers better solutions.

She embraced the mission of NPower after a long career in the corporate sector and with a deeply personal set of motivations. Growing up in a working-class family and the first to graduate from high school, getting a college degree was not a forgone conclusion. It was only through the coaching and counseling of adults who took the time to care that she was set on a very different path, earning a BS in Industrial Engineering and Operations Research at U.C. Berkeley and an MBA from Harvard Business School. It brings her joy to help others connect with their pathway, and to inspire a new generation of leaders to operate at the intersection of good business and better humanity.

Bertina's commitment to helping others break social and economic mobility barriers has led to her involvement in strategic alliance organizations, including founding member of TechPACT; member, CEO Action; steering committee member, American's Promise; Wall Street Journal CEO Council; as well as membership on Forbes Council, NationSwell, and Concordia communities. She was named as one of the Tech Industry's Brightest Superstars by the US Black Engineer & IT magazine. She is an engaged storyteller and speaks frequently at association and industry events and conferences such as SXSW EDU, JFF Horizons, SIM, ASU+GSV, and is a frequent guest commentator across numerous media outlets.

A native Californian, Bertina is a proud resident of Brooklyn where she lives with her husband and teenage son.

 **Susanne Tedrick**—The tech industry wasn't always on Susanne Tedrick's radar. The Northwestern University grad enjoyed a decade-long career in operations and administrative-type roles before craving a new challenge and embarking upon a career change. It was a prescient move. With a critically acclaimed book under her belt, Susanne Tedrick is not only a groundbreaker but one of the leading lights in a rapidly evolving industry.

Her first job in the tech industry, after orchestrating her career turnaround, was as a cloud technical specialist at IBM for which she received a Rising Star of The Year award in 2018 from CompTIA. The subsequent feature in CompTIA World Magazine led to the opportunity to write her first critically acclaimed book, *Women of Color In Tech: A Blueprint for Inspiring and Mentoring the Next Generation of Technology Innovators.* It detailed her personal journey into tech as a Black woman, helped to inform and inspire women of color to pursue tech careers, and made the case that diversity, equity, and inclusion benefits everyone in the technology industry.

Tedrick joined Microsoft as an infrastructure specialist for the Azure for Sports sales team covering all major U.S.-based sports leagues and affiliated teams. She currently works as a technical trainer in the organization, delivering outcomes-based training to Microsoft's leading enterprise customers on its cloud computing platform, Azure.

Susanne has previously been featured in many influential tech and business media outlets including Worth Magazine, CompTIA, PECB Insights, and CIO.com. She has also appeared on numerous podcasts, YouTube interviews, and panel discussions. Her awards and honors include Microsoft's Platinum Club Award, Thinkers360 Top 50 Thought Leader in Cloud Computing, and CompTIA's Diversity in Technology Leadership Award in 2020.

Susanne has taken an active role in community service with several nonprofits including formerly serving as chair of the Advancing Tech Talent and Diversity Executive Council for CompTIA. She is also a coalition member for NPower's Command Shift initiative.

Susanne holds a degree in Communication Systems from Northwestern University and is currently an Executive MBA candidate at New York University's Stern School of Business. Outside of work, Susanne spends time with her wonderful husband Paul, is on a ongoing journey to teach herself electric guitar, and casually indulges in her love of video games.

Acknowledgments

The authors would like to thank:

Our phenomenal research and editorial team who helped to bring this book to life: Lori May, Tracy Brown, and Christine O'Connor.

Two patient and invested individuals who provided outstanding reflections, counsel, and support throughout the process: Viola Maxwell Thomas and Brianne Wilson.

Our marketing, publicity, and social media teams, for sharing our message: Betsy Jones of the Countdown Group, Adrienne Fontaine, David Hahn, and Sharon Farnell at MediaConnect, Keith Simmons of Belay Consulting, and Gogi Randhawa of Generation Esports & Video Game Consulting.

Our extended team at Wiley: Ken Brown, Pete Gaughan, Melissa Burlock, Connor Cairoli, Barath Kumar Rajasekaran, Saravanan Dakshinamurthy, Kim Cofer, and Amy Laudicano.

To Henk van Assen and Wendie Lee who provided expert design recommendations and brilliant illustrations.

To our advisors and contributors, thank you so much for your time and insights, without which this book would not be possible: Frank Baitman, Anita Balaraman, Peter Balis, Jon Beyman, Michael C. Bush, Wendy Myers Cambor, Patrick Cohen, Austin Cole, Byron Cooper, Katty Coulson, Alex Cristan, Cecile Cromartie, Craig Cuffie, Kim Davis, Lena DeLaet, Guillermo Diaz, Stephen Ezeji-Okoye, Iris Fagan, Bob Ferrell, Jill Finlayson, Maria Flynn, Michele Garfinkel, Jill Going, Jeff K. Goldenarrow, Rich Greenbaum, Steve Guiliani, Nell Haslett-Brousse, Jesse Hillman,

Jim Kavanaugh, Helen Kim, Ellie King, Jennifer Kleinert, Danica Lohja, Bobbie Long, Ann Marr, Dan Maslowski, Rasheena McConneaughey, Jacob McConnell, Peter Miscovich, Rebecca Moss, Hunter Muller, Jeff Muti, Andrew Parlock, Marina Perla, Raymond Pitts, Damiete Roberts, Sarah Russell, JT Saunders, Malcolm Smith, Yvette Steele, Dave Steward, Sondra Sutton-Phuong, Roselinde Torres, Marcus Valentine, Robert Vaughn, Miguel Velazco, Pallavi Verma, Kevin Walters, Moja Williams, Matthew Yee, Chris Young, and Flynn Zaiger.

Bertina's acknowledgments:
Most importantly, I am enormously grateful for the partnership, camaraderie, and friendship of Susanne Tedrick. She is a deep thinker, observer, and fearless advocate for fairness and equity in business, and in life.

I am very lucky that my family, especially my husband Brendan Coburn and son John, demonstrated deep wells of patience during the endless evenings and weekends I was in the "book vortex."

I want to thank the entire executive team at NPower: Stefanie Boles, Melody Brown, Kim Mitchell, Kendra Parlock, Tom Sussman, Bea Tassot, Binta Vann, Robert Vaughn, and Felecia Webb for leading the way on DEI in practice. A special heartfelt thanks to Marguerite Durret who somehow magically turns stones into time, especially when time just disappears. And to Rick, Tom, Ryan G, Ryan T, Sergio, Alma, and Henry, thank you for believing I could complete this project!

Susanne's acknowledgments:
First and foremost, thank you to my wonderful writing partner, advocate, and friend, Bertina Ceccarelli. Her passion and tenacity shine throughout and it has been a true privilege working with her.

To my wonderful husband Paul, my dad Ken, my stepmom Jackie, my extended family, and friends—thank you for always cheering me on, as well as for your continued patience, kindness, love, and support. I would not be here without you. A very special thank you to one of my lifelong friends, MariaCriselda Loleng, for helping to amplify our message.

To my most ardent mentors, sponsors, and advocates—Marc Bulandr, Val Haskell, Joanna Vahlsing, Tim Dickson, Dr. Natalie Maggitt, Neferteri Strickland, Linda Calvin, and Kirk Yamatani—your unwavering support and belief in me means more to me than I can ever express adequately. Thank you for always being in my corner.

To my late mother, Susan—your strength inspires me to keep pushing and keep striving always. You are very much loved and missed.

Contents at a Glance

Contents

Foreword

"If it's not personal, you will never achieve your purpose"

The fact that you are reading this makes me hopeful. You are either one of the committed or one of the curious. You are an ally, or you are curious about being an ally. Your journey has begun. Your reward will be a more equitable world where businesses and communities will gain a new untapped competitive advantage and source of positive global contribution. I am confident that *Innovating for Diversity* will be a cornerstone of your journey and the path that you are about to create within your organization.

Reality

A few days ago, I had an "off the record" conversation with DEIB leaders from some of the biggest and most powerful brands on earth. They were tired, and confused, yet committed to changing their organizations. They talked about the need to change jobs or take a break every three years to continue this work. They talked about the fact that burnout was inevitable and coached each other regarding what to do *when* it occurs, not *if* it occurs. I don't know of any other Fortune 1000 C-Suite role where leaders talk this way. They also talked about not wanting to be a DEIB leader who was smiling and "happy to be here," while a quick Google scan of the top 500 leaders showed that their organizations had made very little progress outside of starting a few ERGs and hosting monthly celebrations. We gathered to collaborate, support each other, and double down on our efforts and committed to real change, not marketing change.

Does Your Organization Have the Best Talent?

Organizations do a poor job of measuring performance and therefore can't really know that they have the best people. In fact, the data says that they do not because they are overweighted in a demographic group or two and are underrepresented in many others. This math statistically leads to a normal distribution that yields a suboptimal talent level. As I watch the World Cup soccer matches, I am struck by the brown-skinned people on teams in Europe that were born in their country, speak the language of their country, and are thriving on the soccer field. Their talent is fully embraced. Yet, they look very different from the majority population in their country, and when I do business in their country, I do not see ANYONE who looks like them. I have also yet to see a brown-skinned manager or coach on the pitch.

Are brown-skinned people only good at athletics but not in business or leadership roles?

Not long ago, the NBA (National Basketball Association) was all white. Things have changed drastically and have continued to change. The league was integrated, the level of play increased (they didn't have the best talent with one dominant demographic group), and now it is 30 percent made of players born outside of the U.S. (the U.S. didn't have a monopoly on the best talent after all) and the four MVP candidates in 2021 were all born outside of the U.S. Half of the teams now have Black coaches and I predict half of them will have owners from underrepresented groups by 2025. By the way, teams that were worth hundreds of millions are now worth multiple billions.

So, I believe it is impossible to say that an organization with one or two dominant demographic groups has the best talent. Sure, businesses can do fine with one or two dominant demographic groups. Tons of data support that. But can they be Great? Not much evidence here when we think about our communities, equality, equity, and the environment.

Courage

You have it. You opened the book. You are not afraid. You are an ally. The leaders interviewed for *Innovating for Diversity* have the courage to focus relentlessly on their North Star—that DEIB matters and taking fresh approaches and risks are necessary. They are fearless and unapologetic. They understand that the time to improve DEIB is now—not someday or eventually. They are willing to put in the hard work and cash in on their social capital to make that happen. Now.

Hope

This book gives me hope. Its timing is right globally and impeccably right in the U.S. It is written by two analytical, high-achieving mavericks who have a track record of innovation against all odds. They have been chosen to help implement the change the world needs now. The world needs increased productivity, performance, and innovation while reversing climate change. The challenge should excite you. Collaborators interviewed in this book have a few things in common. They have embraced these challenges by unapologetically applying the growth mindset of Diversity, Equity, Inclusion, and Belonging as the answer. They are also connected by the divine knowledge that the time is now and innovation without execution leads back to the present. The fact is the greatest companies and organizations on earth could be so much greater, and I have the data to prove it.

I will be filled with hope when I see this book on people's bookshelves over their shoulders on video cams. I hope to see a worn or torn cover showing it is well-used. I hope to see two or three copies on the bookshelf because the ally keeps a few extra copies available to give to a curious ally.

Michael C. Bush
Global CEO
Great Place to Work

Introduction: How the Concept of "Innovating for Diversity" Was Born

The origins of this book date back to March 12, 2020, exactly one week prior to the first Stay at Home Order issued in the U.S. as a measure to prevent the spread of COVID-19. It was March 12th that then-Governor Cuomo proclaimed as "Women of Color in Tech Day" across New York State. To commemorate the day, staff from NPower, the sponsoring non-profit for the proclamation, as well as numerous community and corporate partners gathered to ring the closing bell at NASDAQ in Times Square.

One of the guests invited to celebrate the day, Dawn Michelle Hardy, just happened to be the publicist of the author and technologist, Susanne Tedrick. During the event, Dawn introduced herself to NPower CEO Bertina Ceccarelli, and mentioned, "I know the woman who literally wrote the book, *Women of Color in Tech!*".

Susanne's first book had just been published by Wiley earlier that month. Bertina knew then she couldn't pass up the opportunity to meet Susanne and learn about her work and experiences, especially since NPower was seeking expertise on new ways to increase the number of minority women in the organization's IT skills training and job placement program. At the time, less than 4 percent of those working in U.S. tech jobs were Black, Brown, or Indigenous women.

Our first meeting led to numerous reflective and candid conversations about the state of diversity across U.S. businesses, large and small. While we come from two different perspectives, set of lived experiences, and career journeys, we both share common concerns as well as a similar sense of optimism as more inclusive talent management practices take hold in a wide range of industry sectors. Our concerns, like those of some of the leaders we interviewed for this book, stem from a belief that successful efforts to increase diversity, equity, inclusion (DEI), belonging, and accessibility require unwavering commitment and senior executive focus. As economic cycles and labor market supply and demand ebb and flow, history suggests that sustained efforts to improve diversity can slip down the priority list.

Our optimism, however, is fueled by numerous observations, interactions, and interviews with leaders who, frankly, give a damn. That statement may not be common corporate DEI-speak, but we have uncovered that some of the most powerful and effective DEI strategies are instigated by individuals who are fearless about tapping into their own experiences and tackling the root problems preventing inclusive practices in a division or entire company from thriving. They are leaders—sometimes middle managers and sometimes CEOs—who are willing to be creative, take risks, galvanize colleagues, engage human resource counterparts, and pilot new approaches to diversity and inclusion that often inform changes to underlying systems and processes. They are doing anything but simply "checking the box" to advance diversity principles. In short, they are leaders who are *innovating for diversity*.

Why We Wrote This Book

Throughout our own careers over the past three decades in technology and business, we have both witnessed the power of innovation to transform entire industries. We have read the research and seen first-hand how diverse teams catalyze new lines of inquiry and inspire invention. But in recent years, we have met remarkable leaders who are addressing business challenges through innovative solutions that build, develop, and retain diverse teams.

In other words, just as diverse teams drive innovation, innovation principles can be applied to advancing practices that build diverse teams.

We wrote this book to tell the stories of the leaders and companies that are succeeding by applying an innovation lens to diversity. Few of the leaders we interviewed would readily call themselves "DEI innovators." In fact, most would say they were simply doing what they do in their businesses every day: solving problems by addressing root causes. When we deconstructed their solutions—whether apprenticeships or reinventing mentorship—we found they used tools and principles characteristic of innovation. Each set of leaders we profile had the courage to question established practices and put their reputations on the line for what they believed would be dramatic improvements to the status quo.

Something else emerged from our interviews that struck us as worthy of exploration: personal, intrinsic motivation. The individuals at the center of our case studies often shared deeply personal experiences that shaped their own values and beliefs about the need for diverse, inclusive work environments. They exemplify leaders comfortable with vulnerability, humility, and the acknowledgment of others who believed in them and supported their own career objectives. Some define their personal motivation as "paying it forward," others by a sense of purpose or moral imperative. All unquestionably believe richly diverse teams and equitable cultures are, at the core, a business imperative, and produce superior outcomes. None expressed the sentiment that DEI was somehow "someone else's job," primarily an ESG or charitable endeavor, or that it was something that could be "solved" in the short term without persistent commitment. Each was unafraid of accountability.

We also take care to present interviews of those who emerged as full partners following the implementation of practices described in the case studies. In some instances, these are individuals who may not have been hired under prior practices, or who may not have been considered candidates for promotional opportunities. Their voices and stories are essential to the notion of a virtuous cycle between innovation and diversity, and for continual improvement.

We think the focus on individual stories, case studies, and the achievements of dedicated leaders across an organization is especially important in the context of the myriad corporate CEO proclamations in support of diversity announced after the May 2020 murder of George Floyd. As we discuss in Chapter 9, C-suite commitment is necessary, but not sufficient, for DEI principles to become integral to both culture and business operations. What cannot be overlooked is the importance of mid-level managers, in human resources and in business units, to the successful operationalization of specific actions and processes that make DEI a part of everyday practice. And, as we will see in the case studies, the innovations that lead to broadscale adoption are often fueled by the leadership of those mid-level managers.

This is not to say the leaders and companies we profile are perfect. Arguably, it is the willingness to confront imperfections and, as you will learn in Chapter 1, the readiness to challenge Fixed Practices and Fixed Attitudes that fuels breakthroughs in measurably improving recruitment, talent development, and options for career mobility that in turn yields better DEI outcomes. Possessing enough humility to acknowledge the imperfections of any system, product, or business practice is, in fact, a prerequisite for real innovation to flourish.

What we learned from our research and the dozens of interviews with leaders across industries and functions is that the cultural conditions required for innovation to thrive are not unlike those that support diversity, equity, inclusion, and belonging. Values such as courage, risk-taking, collaboration, and trust create an environment where employees are motivated to invent and drive continual improvement. When supported by the right culture and leadership this motivation extends to innovating for diversity. What that culture and leadership looks like is at the heart of the case studies we share.

How to Use This Book

Our objective is to provide inspiration and direction for business leaders reviewing their own DEI practices, perhaps feeling stuck, or just looking for a way to begin. The case studies are structured to present issues and problems that are frequently confronted by companies across industries and organizations of all sizes. While the solutions are specific to each company profiled, we believe they serve as approaches that can be tailored and applied more broadly. Importantly, *how* each team landed at their respective solution through inquiry, testing, iteration, and expansion is especially worth considering.

You will find we intentionally do not offer up an easy checklist of recommended activities. Rather, we lay out an innovation framework that will challenge readers to take a fresh look at their current DEI efforts and help guide the development of new initiatives that can be embedded across an enterprise. At the conclusion of each case study, we assess how both the process and solution were advanced by tapping into components of the innovation framework. We believe the framework and generalizable case studies will be far more useful tools, rather than a checklist with limited utility, for truly tackling DEI as a business imperative.

Because our own experiences and networks are closely connected to the technology sector, many of the examples we share highlight tech companies or divisions, or technology job functions. However, especially given that representation is historically weak in the tech sector, we believe leaders in other industries can readily borrow from the lessons presented.

Similarly, the scope of our case studies focuses primarily on diversity along the dimensions of race, gender, veteran status, and socio-economic and educational background. We acknowledge there are unique considerations for advancing equity and inclusion for those with disabilities, seen and unseen, as well as those who identify as LGBTQ+. At the time of publication, many of the companies we worked with were launching specific initiatives to support greater diversity defined broadly: working mothers and caregivers, white male allies, multi-generational teams, and those who recently immigrated to the U.S. We believe that approaching diversity and inclusion with an innovation lens is the right place to start, independent of the issue.

Before we dive into the innovation framework and the case studies, we believe it is important to first lay some groundwork. In Chapter 1, we provide a brief history of diversity advancement in the U.S. labor force, as well as a summary of demographic trends. While we consider the business case for diversity to be irrefutable, the changing composition of the U.S. population underpins a need for urgency. Real and more rapid progress requires challenging the Fixed Practices and Fixed Attitudes that today hold us back. We describe those practices and attitudes in some detail, some of which you may witness in your own organizations. Our groundwork continues into Chapter 2, where we establish definitions of common terms used throughout the book.

In Chapter 3, we introduce the innovation framework, which details a set of cultural conditions that must be in place for innovation to thrive—whether to address the challenge of improved diversity and inclusion or to tackle any other business challenge. We also present three roadblocks of those we interviewed repeatedly surfaced as "toxins to innovation": lack of prioritization, inertia, and arrogance. Companies that innovate for diversity and, in turn, leverage diversity for greater innovation, benefit from a virtuous cycle that can help them outperform their competitors.

Chapters 4 through 8 dive into the case studies and the stories of individual leaders and teams that reflect innovating for diversity in action. We have grouped these studies by topics of keen interest to many of the leaders we interviewed: apprenticeships, mentorship, advancing diversity in resource-constrained environments, talent acquisition, partnerships, and talent development and advancement. At the end of these chapters, we provide a summation of the innovation principles that were applied within the case, as well as what steps each company took to mitigate potential threats to innovation in DEI. The case studies in Chapter 9 pay special attention to the role of C-suite leadership in prioritizing diversity and in operationalizing practices that define it as a business necessity, not a passing trend. Our final chapter summarizes lessons and takeaways.

We think you will be as inspired as we are by the leaders you will meet in this book and hope you will take away any number of ideas that are practical and actionable for your business. If after reading the following chapters, you are resolved to be more courageous, more open to risk and collaboration, more committed to building trusting partnerships, and more ready to lead fearlessly when innovating for diversity, then we will have succeeded.

Why Are We (Still) Here?

Today, our country is more richly diverse than ever before. Yet, data shows that even 70-plus years after the passage of legislation to end discriminatory employment practices and significant investment in diversity, equity, and inclusion (DEI) programs, the United States labor workforce still does not reflect the diversity of the population as a whole. In fact, for certain demographic groups, their presence in the workforce has decreased in recent years.

Why have we not moved the needle more significantly in the last seven decades? What are the causes behind this? And where should we go from here? In this chapter, we will review current labor statistics and examine where change has (or hasn't) happened. We will also examine the underlying factors contributing to this stagnation: *Fixed Attitudes, Fixed Practices,* and seismic global events that have left long-lasting repercussions.

Key Concept: Limiting beliefs and ineffective systems and practices have contributed to the persistent lack of diversity, equity, and inclusion across many industries.

A Brief History of Diversity in the U.S. Labor Workforce

The United States is, at the time of this writing, 245 years old, and is a nation founded and sustained by diverse populations. It's hard to remember that our labor workforce did not always reflect our population, and it required years of painstakingly hard work and sacrifice to achieve progress.

One of the very first pieces of legislation to promote diversity in the workforce was in 1948, when President Harry Truman officially desegregated the armed forces with Executive Order 9981. The order made discrimination based on "race, color, religion or national origin" illegal for all members of the armed services and was a crucial victory in the Civil Rights Movement.

The next major piece of legislation was Title VII of the Civil Rights Act of 1964, which prohibited discrimination based on race, color, religion, sex, and national origin when making decisions regarding hiring, promotion, discharge, pay, benefits, training, classification, referral, and other aspects of employment. The passing of Title VII dramatically impacted the U.S. labor workforce, sharply increasing the labor force participation of women and people of color. It was amended over time to prohibit discrimination on the basis of sexual orientation and gender identity.

In tandem with Title VII, the Equal Employment Opportunity Commission (EEOC) was also created to enforce the provisions of Title VII. At the time of its creation, the EEOC's powers were fairly limited to only investigating claims of discrimination, but not to compel employers to comply with Title VII. That changed in 1972 with the Equal Employment Opportunity Act, which gave the EEOC the authority to file lawsuits against private companies to impose compliance.

The Age Discrimination Act (ADEA) passed in 1967, and prohibited discriminatory hiring practices against those aged 40 and older, while the Americans with Disabilities Act (ADA) of 1990 outlawed discrimination of disabilities. Finally, the Lilly Ledbetter Fair Pay Act of 2009 was an important piece of legislation to ensure equitable pay, regardless of gender.

The Current United States Labor Workforce

Let's now look at the current U.S. workforce. As shown in Table 1.1, in February 2020, at the start of the global COVID-19 pandemic, the total

United States labor workforce was made up of approximately 164 million people, according to the Bureau of Labor Statistics.[1] Due to the effects of the COVID-19 pandemic, the labor workforce experienced a sharp drop, but was on the rebound beginning in February 2022. Compared to 2000, this represented a 15 percent increase.

Table 1.1: United States labor workforce in 2000, 2010, 2020, and 2022

PERIOD	TOTAL WORKFORCE (000)
February 2022	163,991
February 2020	164,583
February 2010	153,512
February 2000	142,456

Examining Labor Workforce Dimensions

Within the context of the protections put in place to promote diversity, persistent disparities in employment representation and income are at best surprising, and at worst, alarming.

For this exercise we reviewed data in several areas from 2000, 2010, and 2020. Figure 1.1 and Figure 1.2 examine representation of the working-age population and the labor workforce by race and gender, as well as income differentials per the U.S. Bureau of Labor Statistics (BLS) and the U.S. Census Bureau.

The data reveals that the aggregate labor force continues to be over-represented by white workers and underrepresented by Black workers over the past two decades. As of 2020, working-age Hispanics partic-ipated in the labor force at a level equivalent to their numbers in the 16–64 population group, as has been the case for Asians consistently over the past 20 years. More startling is the continued income disparity between Black and Hispanic employees and their white counterparts. While we've seen modest progress over the past 20 years, especially for Hispanic workers, both Blacks and Hispanics earn, on average, less than 80 cents for every dollar earned by white workers. This is, in part, a reflection of Black and Hispanic workers disproportionately represented

---

[1]The labor workforce is defined as all workers, 16 years of age and older, who are available for work. This includes workers who are either employed or unemployed, and are not of active status in the United States Military.

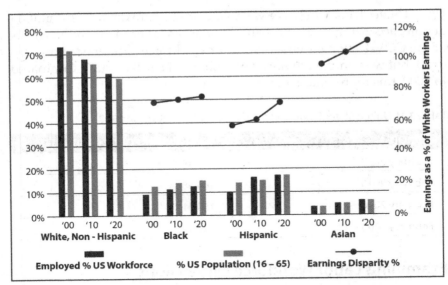

Figure 1.1:   United States employment and income disparity by race

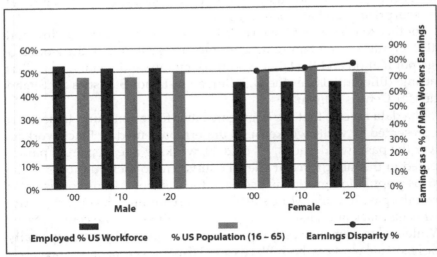

Figure 1.2:   United States employment and income disparity by gender

in frontline, lower-wage jobs and less so in managerial positions. Women have similarly continued to lag in their wages, earning on average 82 cents for every dollar earned by all men in 2020, making slight gains in a 20-year period. For women of color, the differential is even more stark: Latinas earned on average 49 cents and Black women 58 cents for every dollar earned by white, non-Hispanic men in 2020. This compares to 73 cents earned by white women compared to white men.[2]

Some of the income differentials can be explained by educational attainment. Figure 1.3 and Figure 1.4 show educational attainment by race and gender during 2003, 2010, and 2020. Overall, Asians have the highest completion rates of four-year and advanced degrees, which in part drives average income levels that are 112 percent of their white counterparts. For Blacks and Hispanics, educational trends are similar to those for income: both groups are making gains in educational attainment but still complete four-year and advanced degrees at levels between 40 percent and 70 percent of white adults.

While education is one factor contributing to unequal earning potential, persistent racial and gender discrimination impacts access to stable jobs with good wages, health benefits, and retirement plans.

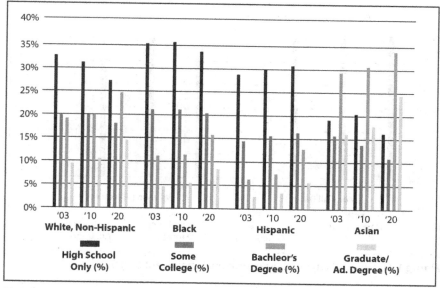

Figure 1.3:   United States highest educational attainment by race

Lower earning potential combined with historic discriminatory practices that have reduced access to mortgages and tax-advantaged forms of savings have widened the wealth gap between white households and

---

[2]National Partnership for Women & Families. "Quantifying America's Gender Wage Gap," May 2022.

Black and Hispanic households. A 2021 report by the St. Louis Federal Reserve Bank showed that the median wealth owned by the average white family was $184,000 compared to $23,000 for Black families and $38,000 for Hispanic families. The gap for Black families was largely unchanged in 30 years and only marginally improved for Hispanic families. Of note, wealth gaps persist regardless of education level. For example, Black and Hispanic households where both parents have a bachelor's degree have an average family wealth at the 40th and 49th percentile, compared to white households with the same educational background that own wealth at the 65th percentile. This wealth disparity makes it more challenging for Black and Hispanic parents to support their children in paying for higher education, buying a home, or transferring significant resources from generation to generation.

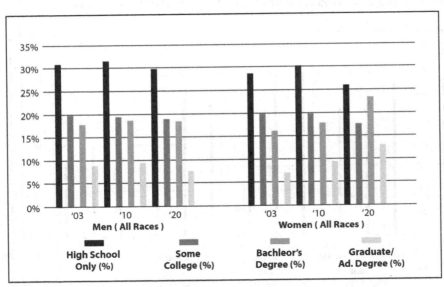

Figure 1.4:   United States highest educational attainment by gender

On the major dimensions just discussed, our progress in improving representation and income across race and gender in a two-decade time frame has been slow, even with new investment and the additional awareness of the benefits and need for diversity.

It should also be noted that even the collection of diverse labor workforce data by the BLS is limited. For example, the data available on these dimensions prior to 1965 is scarce. It also doesn't account for neurodiverse populations, workers with physical disabilities, or workers who identify as part of the LGBTQ+ community.

Other statistics of note:

- For management occupations, Black, Hispanic, and Asian American/ Pacific Islander (AAPI) workers held 8.9 percent, 10.8 percent, and 7.1 percent, respectively of these jobs in 2019. Women held 38 percent of these professions.

- Looking at Fortune 500 CEOs, in 2021, only 4 were Black, and only 41 were women.

- Only 23 percent of women have C-suite positions. Of that, women of color account for only 4 percent of that figure.

- Hispanic workers are overrepresented in lower-wage occupations such as painters, construction labor, and housekeeping cleaners, while Black workers over-index in occupations such as security guards, home health aides, and bus drivers.

- A large share of employed women across all race and ethnicity groups work in lower average wage occupations connected to education and health services: Black women (41 percent), white women (37 percent), Asian women (32 percent), and Hispanic women (31 percent).

Unfortunately, there are many more statistics that point to opportunity gaps and disparities that continue to exist along the continuum between educational attainment, employment, and household wealth. At a high level, too many efforts to reverse these disparities, while well-intentioned, produce underwhelming results.

The Murder of George Floyd, the Rise of the Black Lives Matter Movement, and the Corporate Response

There have certainly been a few notable social justice movements aimed at pushing companies and other institutions to evaluate how they address pervasive societal issues. In recent years, the MeToo movement gave organizations a moment to reflect on whether they were doing enough to protect their female employees from harassment (blatant and subtle), while the Occupy Wall Street movement highlighted persistent income inequality across the U.S. But perhaps no one movement in recent history has garnered as much attention and as strong a response from companies, think tanks, and governmental entities than the Black Lives Matter (BLM) movement.

Started in 2013, after the acquittal of George Zimmerman—who killed Trayvon Martin, an unarmed Black teenager—the BLM movement is considered to be a loosely structured social movement, aimed at revealing to wider audiences the struggles and inequities Black people face on a daily basis. The movement also seeks social justice reform, including how Black people are treated by law enforcement in comparison to other racial and ethnic groups.

The movement's prominence was elevated significantly in 2020, with the murder of George Floyd at the hands of white police officers, chiefly Derek Chauvin, in Minnesota. In response, protests across the country in May 2020 gave rise to civil unrest and moral outrage.

Many employers struggled with how exactly they should respond, if at all. Within workplaces the civil unrest sparked critical discussions on what employers owe their employees of color to address social justice and stand up as a corporate citizen responsive to the moment. For most, it was walking a tightrope—don't respond and you're labeled as uncaring and an employer who perpetuates the very social inequities BLM speaks out against. Respond with the wrong message, and you risk amplification of negative perceptions.

We also witnessed many employers spending liberally on traditional and social media to decry what occurred. But for many people of color, perhaps because of a weariness stemming from years of empty promises and forgotten commitments, these proclamations did nothing but confirm their belief that the plethora of corporate statements were all for optics, with no meaningful change forthcoming.

Some employers took a step further to stake a quantitative commitment to change. This was in the form of an increased focus on diverse hiring, diversity training, forming listening spaces where people of color could share and process their experiences with others, and/or making investments in Black-owned businesses. Using the tech industry as an example, the top 42 tech companies pledged nearly $4 billion in further developing their DEI programs, investing in Black-owned companies and banks, and making bolder commitments to bringing in more diverse candidates.

Fast-forward to a year later, many wondered—and rightfully so—has corporate America made good on its promises? The honest answer: It depends on whom you ask. It cannot be denied that conversations on race and social justice within organizations were happening more frequently. Per a study from the *Financial Times* the term "social justice" was mentioned over 300 times and the reference to "Black Lives Matter" was used nearly 150 times during corporate earnings calls of publicly traded companies within the first calendar quarter of 2021. In a similar study by Bloomberg, the mention of terms "racism," "equality," and "social justice" during S&P 500 company earnings calls reached an all-time high in the third quarter of 2020.

But it also can't be denied that little progress has been made in the two years after George Floyd's murder. Workforce diversity metrics have not moved significantly, as we've discussed previously. From the previous statistic on mentions of "social justice" and "Black Lives Matter," those terms were mentioned much less frequently by the start of the second quarter 2021. For perspective, there are over 2,000 companies that are publicly listed with the New York Stock Exchange, and nearly 4,000 are listed on NASDAQ. With that backdrop, the numbers of these companies having frank conversations on race and social justice remain stubbornly low.

Why Haven't We Made More Progress?

With so many companies increasingly vocal and visible in acknowledging that good DEI practices advance the bottom line—and are important legally and morally—why do we still see glaring gaps in representation in high-growth sectors of our economy? Issues underlying our collectively sluggish progress include a continuing reliance on entrenched formulas (Fixed Practices) and a pervasiveness of ingrained behaviors and resistant cultures (Fixed Attitudes) that are not responsive to social and demographic dynamics colliding in today's U.S. workplace.

First, America has never been a more diverse country, ranked along dimensions of race, ethnicity, age, gender, and sexual identification. The U.S. Census Bureau launched its first "Diversity Index" with the 2020 census to measure the probability that two people chosen at random will be from different race and ethnicity groups. A "0" value indicates everyone in a population has exactly the same race and ethnicity characteristics and a "1" value indicates that everyone in the population has different race and ethnicity characteristics. In 2020 the Diversity Index tells us there is a 61.1 percent probability that two people selected at random will have different characteristics, up significantly from 54.9 percent in 2010, as shown in Figure 1.5. What's even more interesting is how the Index has changed over time by state, with almost every region of the country becoming more racially and ethnically diverse. The racial diversification of the nation is projected to continue, with white Americans falling below 50 percent for the first time around 2045, when there will be no racial majority in the country. William Frey, a demographer at Brookings Institution, projects that by 2060, the Hispanic and Asian populations will roughly double in size, and the multiracial population could triple due to immigration and births.

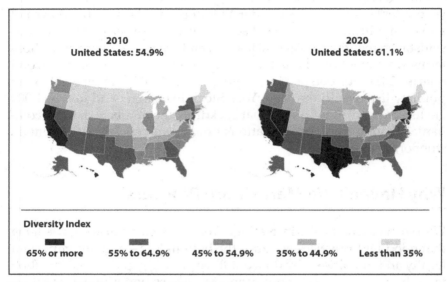

Figure 1.5:   U.S. Census Bureau Diversity Index

The 2020 census also reveals we are living in the most age-diverse era of U.S. history. Compared to 1900, age distribution is far more uniform with almost equal numbers of people from birth to early 70s, as shown in Figure 1.6. One in four people is under 20 years old, and similarly, roughly one in four people are over 60. Sasha Johfre, a sociology researcher at Stanford University, observes that this new era of age diversity offers "an unprecedented opportunity to foster relationships between people of very different ages at a scale never seen before." With five generations together for the first time in the workplace, we can benefit from the new ideas and levels of empathy that intergenerational connections can foster.

Over the past ten years, we have also seen an increase in expressions of gender and sexual diversity. A 2021 Gallup Poll shows that lesbian, gay, bisexual, or transgender (LGBT) identification has increased to 5.6 percent of U.S. adults, as shown in Figure 1.7. Notably, one in six Generation Z respondents (aged 18–23) identified as LGBT, or 16.7 percent, suggesting shifts in both sexual orientation as well as younger adults more willing than their older counterparts to acknowledge their LGBT identity.

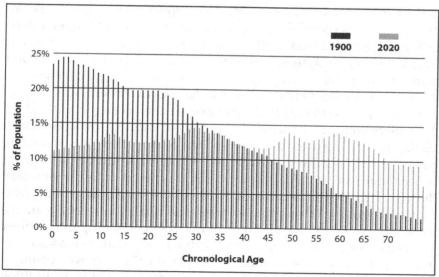

Figure 1.6: Age distribution of U.S. population, 0–74 years old

© 2022 Sasha Johfre. Reprinted with permission.[3]

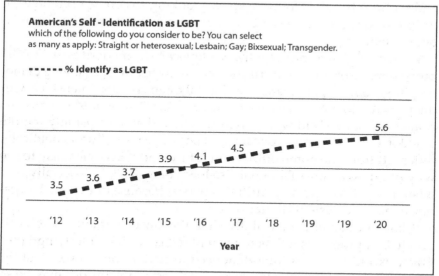

Figure 1.7: LGBT identification

Second, just as the U.S. is becoming more diverse, we have also become a more divisive and polarized society, with both the COVID pandemic

---

[3]Johfre, S. (2022, April 25). In a world experiencing unprecedented age diversity, how should we think about age? Encore.org, from https://encore.org/in-a-world-experiencing-unprecedented-age-diversity-how-should-we-think-about-age/amp

and use of social media accelerating a movement of self-organizing along cultural and ideological views. A 2019 study of social fragmentation in the U.S. by Leila Hedayatifar and her fellow researchers at the New England Complex Systems Institute notes how digital social networks across regional clusters amplify and reinforce viewpoints of like-minded individuals. Rather than eliminating ideological borders, social media communications strengthen them as people interact largely with others who share common views and avoid those with opposing opinions. Hedayatifar's research also references a related trend in self-sorting behavior with people physically moving to communities with other like-minded individuals. This migration was first detailed by journalist Bill Bishop in his influential book *The Big Sort: Why the Clustering of Like-Minded America is Tearing Us Apart* (Mariner, 2009). In recent years the trend has grown as blue and red counties deepen their political hues. The real estate technology company Redfin reported an increase following the pandemic in the number of people looking to move outside of their metro area as remote working became more common. In addition to housing affordability, high on the list of considerations for new destinations was a community with values that offered a fit with political and cultural beliefs. Not only are we looking to connect online with people who share our perspectives, we also want them to be our next-door neighbors.

What does all this mean for the workplace? The modern post-COVID professional workplace is a crucible where the reality of changing demographics, political polarization, and social fragmentation intensifies with the pressure to meet quarterly business goals. Throw into the mix the challenges of a hybrid work environment and an increasingly remote workforce of communities distant (and possibly culturally and politically different) from the home office community, and it becomes easy to see why progress against DEI goals is slow going. This is especially true when Fixed Practices and Attitudes persist because of personal, organizational, and systems limitations.

While in Chapter 2 we will dive into the most common reasons corporate DEI programs fail, here we will focus on the underlying Fixed Practices and Fixed Attitudes that need to be disrupted if we want the pace of change to accelerate. These Practices and Attitudes may be so deeply embedded that they practically disappear into the folds of a corporate culture. These forces are both subtle and non-so-subtle reasons new approaches and ideas flail.

Fixed Practices: Reluctance to Let Go of Entrenched Formulas

Practices for recruiting, on-boarding, developing, retaining, evaluating, and promoting talent are often deeply rooted. They are embedded in a company's way of doing business, organized in playbooks, and shared as best practices. They are further codified through human resources (HR) systems, including software technology and training, that can take years and significant resources to change, especially at scale. Even as the pandemic required rapid shifts in many long-standing business practices—from supply chain management and service delivery to travel and communications—many HR practices remained constant.

Our observation of what is at the root of Fixed Practices, consciously or not, is summarized in Table 1.2:

Table 1.2: Fixed Practices

FIXED PRACTICES RELUCTANCE TO LET GO OF ENTRENCHED FORMULAS	
●—	A declaration and genuine belief by senior executives that "change takes a long time," relinquishing a sense of urgency that signals DEI is not a sincere business priority.
●—	A mental model that assumes a steady stream of well-intentioned activities, such as unconscious bias training or diversity month celebrations, will alone over time change attitudes and behaviors.
●—	A general appreciation of a "business case" for DEI, but without the discipline to apply the same rigorous metrics of outcomes and performance—or resource investment—that would be applied to a new product launch, acquisition, or technology development.
●—	An expectation that a competent HR professional or chief diversity officer will "solve the problem," without recognizing the comprehensive systems and structural changes required of all leaders across an organization. And, dismissing that lead professional when the "problem" isn't solved.
●—	A reluctance to reduce focus on approaches that may be personal to the CEO or senior leaders, like limiting recruitment of entry level talent from the alma maters of executives, or preferencing children of social networks for internships.

Continues

Table 1.2 (*continued*)

FIXED PRACTICES **RELUCTANCE TO LET GO OF ENTRENCHED FORMULAS**	
●———	An interest in propagating a culture of scarcity and prestige by only recruiting candidates with a minimum GPA, inflating skills requirements in job descriptions, or requiring extensive applicant testing for aptitudes or skills not needed for the job.
●———	Legacy or embedded systems and processes that perpetuate overt or subtle bias, such as limited options for racial or gender identity on applications, or algorithm-based hiring systems that reinforce past patterns.

Fixed Attitudes: Continued Pervasiveness of Ingrained Personal Ideas and Beliefs

Our attitudes—or as Wikipedia puts it, our settled ways of thinking or feeling about someone or something—are driving many of reasons we see these Fixed Attitudes continue, even when there is evidence to the contrary that they do not work. The resistance to change when it comes to Fixed Attitudes is often connected to one (or more) of the core emotional responses summarized in Table 1.3 and illustrated in Figure 1.8.

Table 1.3: Fixed Attitudes

FIXED ATTITUDES **CONTINUED PERVASIVENESS OF INGRAINED PERSONAL IDEAS AND BELIEFS**	
Fear	
●———	It's incredibly difficult for leaders and professionals to be vulnerable enough to admit that on DEI matters, they don't have all the answers. Taking that further, there is a fear that if they do or say something wrong, they (and their organizations) will be viewed negatively and reputationally, unable to recover or "canceled." This causes people to completely "check out" or increase their defensiveness when the subject of DEI comes up.
●———	There is also a fear that in some respects, diverse candidates may perform at a less competitive level than their non-diverse counterparts. Their projects and work performance will suffer, so why take on that unnecessary uncertainty?

Continues

FIXED ATTITUDES CONTINUED PERVASIVENESS OF INGRAINED PERSONAL IDEAS AND BELIEFS	
Indifference	
●———	For leaders and professionals who are already successful now, especially from a financial standpoint, they do not see the sense in changing things up for the sake of diversity. To some extent, indifference is driven by a short-term mindset that expects conditions allowing for success today will always exist.
Denial	
●———	This can appear in several ways. Denial can manifest in a belief that there are not enough diverse, and good, candidates in the pipeline. This can also present itself in the belief that cultural biases do not exist within the organization, and that any barriers diverse candidates face in fitting or being successful are largely due to their own shortcomings. Leaders and professional may also be in complete denial that systemic inequities exist.
Anger	
●———	There are some who believe that by investing in these programs and making more room for diverse candidates, that future opportunities are being taken away from them, or worse, they'll be displaced from their job.

The COVID-19 Pandemic: DEI Response to Long-term Structural Impacts

The impacts of COVID-19 on the U.S. workplace, economy, and society will reverberate for years, if not decades. The pandemic challenged companies to rapidly adapt a new social contract with employees as physical and mental health became unexpected priorities, as did the need for increased flexibility and security. Trust took center stage as employees needed safe working conditions and employers required productivity, often from a distance and without direct supervision. Many companies succeeded in reframing their commitment to employees and retained workers, while others experienced a wave of departures. Pew Research

Center found that the "quit rate" reached a 20-year high in November 2021. Its February 2022 survey found the top three reasons American workers cited for quitting their jobs in 2021 were low pay (63 percent), no opportunities for advancement (63 percent), and feeling disrespected at work (57 percent).

Across industry sectors, U.S. workers reassessed their relationship to work, including job satisfaction, purpose, work-life balance, and compensation. This clear-eyed view of the personal meaning of work, especially as millions of Americans confronted death and serious illness of friends and family members, led to the Great Resignation, or Great Reshuffling. Yet, among the wave of people who voluntarily chose to leave their jobs, millions more involuntarily lost their jobs because of the economic downturn. With pandemic shut-downs forcing businesses in the travel, entertainment, and hospitality sectors to close for extended periods, low-wage and part-time workers were especially hard hit.

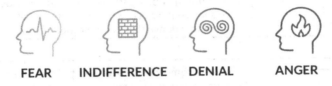

FEAR   INDIFFERENCE   DENIAL   ANGER

Figure 1.8: Fixed Attitudes, Illustrated

Unemployment among Black and Hispanic workers peaked in May 2020 at rates 18–30 percent higher than white workers and remained considerably higher through 2021. Layoffs early in the pandemic severely impacted women of color, especially Black and Latina women. From January 2020 to January 2021 labor force participation among Black and Latina women dropped by 4.5 percent while that for white women dropped 1.5 percent, and by early 2022 a wide gap continued to persist. The disparity in unemployment and labor participation rates for women of color compared to their white counterparts is in part explained by the type of jobs women had prior to the pandemic, as well as the presence of children and elderly parents in the household. Latina and Black women were more likely to exit the workforce to care for young children at home during school closures.

Another group unevenly, and perhaps unexpectedly, impacted by the pandemic were U.S. military veterans and their spouses. This is a population group we will return to in later chapters, with a special focus on enlisted, non-officer service members and active-duty members of the National Guard and Reserves. These are groups that in many workplaces do not widely identify their status to employers, yet through the pandemic experienced high levels of mental health distress, concerns about access to healthcare, and both economic and housing instability often associated with high levels of spousal unemployment.

Taken together, labor and economic statistics reveal that racial minorities and low-income Americans lost substantial financial ground from 2020 to early 2022, further widening wealth, income, and opportunity gaps. Adding to this divide is the longer-term impact to the future workforce of uneven educational outcomes for marginalized populations throughout the pandemic. In June 2021 the U.S. Department of Education reported several striking disparities in the experiences of both public K–12 and postsecondary students from minority populations, those who identify as LGBTQ+, and students with disabilities. Among the observations:

- COVID-19 deepened the impact of disparities in access and opportunity facing students of color in public schools, including technological and other barriers that make it harder to stay engaged in virtual classrooms.

- For many elementary and secondary school students with disabilities, COVID-19 significantly disrupted the services needed to support their academic progress and prevent regression. These disruptions may exacerbate long-standing disability-based disparities in academic achievement.

- Nearly all students (K–12 and postsecondary) experienced some challenges to their mental health and well-being during the pandemic and many lost access to school-based services and supports, with early research showing disparities based on race, ethnicity, LGBTQ+ identity, and other factors.

- Heightened risks of sexual harassment, abuse, and violence during the pandemic, including from household members and online harassment from peers and others, affect many students and have a continued disparate impact on K–12 and postsecondary girls and women and students who are transgender, non-binary, or gender non-conforming.

- Identity-based violence could have long harmful effects on targeted students and their communities. Since the beginning of the pandemic, Asian American and Pacific Islander students in particular faced increased risk of harassment, discrimination, and other harms that affected their access to educational opportunities.

Many institutions of higher education that disproportionately serve students of color and students from low-income backgrounds saw declines in enrollment throughout the pandemic. During the 2020–21 academic year Historically Black Colleges and Universities (HBCUs), Minority Serving Institutions (MSIs), and Tribal Colleges and Universities (TCUs) had declines in enrollment that in some cases far outpaced enrollment declines in their predominantly white peer institutions. Similarly, students in higher education with disabilities faced significant barriers during COVID-19, such as reduced access to academic accommodates and limited communications with faculty, that threatened their educational advancement.

The short- and long-term consequences of the pandemic on employment, income, and educational outcomes—especially among minority and marginalized communities—add even greater urgency and relevance to DEI progress at American corporations. Populations disproportionately affected are both consumers and current and future employees. Government policies that mitigate employment and educational setbacks through investment in training, community colleges, access to childcare, services for those with disabilities, and unemployment benefits certainly help. But smart and persistent efforts by companies to attract, train, retain, and develop talent from across groups underrepresented in their current ranks will provide the longer-term benefits of a larger labor pool together with the macroeconomic benefits of household income gains and robust growth in GDP. And, to return to a point we mention at the beginning of this section, a focus on creating more inclusive, respectful cultures with room to grow will also attract those who voluntarily quit their companies out of sheer frustration.

Yes, a relentless focus on DEI at all levels of an organization is not only the right thing to do, but the economic and cultural benefits for the company and country are undeniable. This brings us to the business, ethical, and moral case for approaching DEI goals with the same serious intent as any business goal corporate leaders signal to investors and other stakeholders.

Why Diversity Matters

We can all agree that having diverse voices at the table yields rich and fascinating interactions, as well as opportunities to learn from one another. But because DEI programs are often an HR-led function, it is routinely viewed as a cost center, or a function that does not add to the bottom line of the organization. This view of DEI is narrow minded and misses a much broader opportunity.

The Business Case for Diversity

Numerous studies in recent years have documented the business imperative fostering a high-functioning diverse and inclusive workforce. Highlighting a few findings, companies with strong DEI initiatives are:

- More likely to meet or *exceed* their financial targets
- More likely to have high-performing teams
- More likely to be considered more agile and innovative, and
- More likely to achieve or exceed their business outcomes

A comprehensive longitudinal study by McKinsey & Company looking at data from hundreds of companies from 2014–2019 found that workplaces with strong gender and ethnic diversity are more likely to produce better financial results than their less diverse competitors. The study found that in 2019 companies in the top quartile for gender diversity on executive teams were *25 percent more likely to have above-average profitability* than companies in the bottom quartile, which was up from 21 percent in 2017 and 15 percent in 2014. In the case of ethnic and cultural diversity, top-quartile companies outperformed those in the bottom quartile by *36 percent in profitability* in 2019.

But the business case goes beyond quarterly financial results. As competition for talent intensifies, companies viewed as prioritizing diversity and inclusion are better positioned to win. The popular job search engine Glassdoor reports that 67 percent of job seekers say that having a diverse environment was a key factor in their decision to work for an employer. Like employees, the investment community is also prioritizing DEI criteria. Over 180 private equity investment groups, such as KKR, Warburg Pincus, and Vista Equity Partners have committed to the Institutional Limited Partners Association's (ILPA) Diversity in Action initiative launched in December 2020. The Diversity in Action initiative requires participants to track hiring and promotions by race and gender and report employee demographic data while raising funds. Private equity firms low in DEI indicators risk losing coveted investments from large-scale institutional investors.

In Chapter 3 we will explore in more detail the specific connection between diversity and the financial results generated from innovation.

While the business case is often the most expedient short-hand for explaining the value of DEI to senior leaders, it is imperfect. The risk of over-promoting the promise of improved financial performance is that it can narrow DEI initiatives to transactional relationships. A heavy emphasis on recruitment numbers without equal consideration to inclusivity can cause underrepresented populations to feel they are in a position of having to constantly justify their presence and consistently outperform their majority counterparts, undermining any hope for real equity.

The Moral and Ethical Imperative for Diversity

We agree that DEI efforts, when executed well, can improve a business's financial bottom line. Our opinion, though, is that these business drivers should not be the *only* reason to act. It cannot be stressed enough that there is a compelling moral and ethical case to pursue DEI programs. All workers deserve the right not only to be treated with dignity and compensated equitably, but to have access to employment that fully engages them, and full agency over their career path.

We must also not take for granted that if we are not vigilant in defending the need for DEI, we may begin to witness a roll back of many decades of progress in workforce diversity. As mentioned earlier, our society has become polarized, with DEI opponents pushing back on social justice policies designed to advance diversity. As one example, in March 2022, California's landmark law requiring corporations to diversify their boards with representation along race, ethnicity, gender, and sexual orientation was struck down as being unconstitutional.

Another, more concerning example: in Florida, House Bill 7/Senate Bill 148 (HB7), also referred to as the "Stop Woke Act" went into effect in July 2022, effectively prohibiting certain employers from delivering DEI training that uses critical race theory, or subject matter that may make certain groups feel uncomfortable, as it may suggest that they are to blame for past transgressions against historically marginalized groups.

We are not supporters of "blame and shame"-oriented DEI programs (more on that in Chapter 2) but avoiding the hard truths and conversations about our history and our *continued* systemic inequities does not give us an opportunity to learn from our past and increases our likelihood of repeating these mistakes in the future. Additionally, DEI programs that attempt to minimize these aspects, and only amplify positivity and progress without acknowledging the work still to be done, can make diverse talent feel as if they are being gaslit by those around them. Well-intentioned leaders who urge colleagues to "be patient" or proclaim "you're too sensitive" when it comes to DEI, undermine the very progress they want to make.

What Got Us Here Won't Get Us There: The Diversity-Innovation Paradigm

We began this chapter with the question, "Why are we still here?" While there has certainly been progress at the individual corporate and industry level, workforce and attitudinal data show we have a long road ahead to build cultures that prioritize and benefit from diverse, inclusive, and equitable work environments. To make the dramatic leap forward required for companies to meet ambitious long-term goals announced in a season of social and racial unrest—and simply to respond to the changing demographic, cultural, and postpandemic conditions in the U.S.—the Fixed Practices and Fixed Attitudes that have long prevailed need to be identified and disrupted. We believe there is immense opportunity for innovation at every level of DEI practice to loosen rigid procedures and mindsets that no longer work.

In Chapter 3, we look at what it takes for innovation to flourish for any purpose, in any industry. We review barriers that often impede innovation: lack of prioritization, inertia, and arrogance. We also consider five cultural conditions required for innovation: courage, risk-taking, trust, collaboration, and leadership. When applied to DEI, we begin to see how Fixed Practices and Fixed Attitudes can in fact be shifted, making way for fresh thinking and new pathways to real and lasting progress.

What is motivating and promising to both of us is how smart companies and leaders are innovating today in ways both large and small. We have had the privilege to work with extraordinary leaders who are deeply committed to building inclusive cultures and practices. They are individuals who often risked their own professional reputations or challenged ingrained practices to pursue new approaches to attracting and promoting talent or more effectively unlocking the diverse perspectives on their current teams. Not satisfied with just making progress in their own areas, the most successful leaders found ways to operationalize their innovation to advance results across the entire company. These are the stories we are excited to share in the case studies in Chapters 4–9.

Conclusion

In Chapter 1, we examined the growth and change of the United States labor workforce over the past 20 years. While DEI laws and practices have evolved over the past seven decades promoting a marked improvement in workforce representation, current statistics and seismic world events have made it abundantly clear that disparities still exist. And although the reasons for these disparities are myriad, at the root is a reluctance and unwillingness to let go of ill-suited practices and beliefs.

In Chapter 2, we will explore, in greater detail, what the terms *diversity*, *equity*, and *inclusion* really mean, and how each is needed in order to create meaningful change in an organization.

Summary

- Despite the passage of strict labor laws and the rise of adoption of DEI programs, diverse representation and income parity is still lacking in the United States labor workforce.

- Events such as the COVID-19 pandemic and the murder of George Floyd have forced leaders in organizations to examine their DEI efforts.

- The reluctance of organizations to let go of entrenched formulas (Fixed Practices) and ingrained personal ideas and beliefs (Fixed Attitudes) are the root causes of the stagnation of workforce diversity.

- Aside from the moral and ethical imperative, diversity yields tangible business benefits for *all organizations*.

- Leveraging principles of innovation can greatly help leaders and organizations in developing meaningful and impactful DEI programs.

Defining Diversity, Equity, and Inclusion

In some respects, the phrase *diversity, equity, and inclusion* (DEI) has become a buzzword. The terms within the phrase have also been used interchangeably. However, each term has specific meanings, distinctions, and implications.

In this chapter, we will examine and explore each term on its own, how they are connected to one another, and how each is needed to create meaningful change. We'll also examine, at a high level, why organizational DEI efforts often flounder or fail.

Key Concept: Diversity, equity, and inclusion are individual elements that need to be understood fully, and must be used together, for any initiatives to be successful.

What Do Diversity, Equity, and Inclusion *Actually* Mean?

The term DEI has risen in prominence significantly in the past decade. Although workplace diversity training programs have been around since the 1960s (developed in response to the passage of affirmative action and equal employment opportunity laws), there has been more focus on the subject as our society continues to examine and rectify years of social injustice against historically marginalized populations.

Despite the concept being around for some time, people and organizations still struggle with defining what exactly DEI entails beyond surface-level explanations. Organizations may use the term to describe their efforts, but may execute on only one or two dimensions or interpret that each of the expressions can be used in place of the other.

It is important to have a firm grasp not only of what *diversity*, *equity*, and *inclusion* mean, but to understand what each looks like in practice. Lack of this baseline understanding can lead to misalignment on the work needed.

What Is Diversity?

Diversity can be defined as any dimension where one group of people can be differentiated from another. In most conversations on diversity in the corporate context, the focus is primarily on promoting gender and racial diversity. But diversity can also encompass sexual orientation, age, ethnicity, social class, and disabilities (physical, neurological, visual, auditory), among others.

The beauty of diversity is the variety that is achieved by bringing together people from different backgrounds and experiences. Diversity should not be viewed as a mere tolerance of differences, but a genuine appreciation and respect for these differences.

Ellie King, co-founder of Equal IT, a UK-based mission-driven business working to diversify tech teams, says diversity is "about bringing everybody to the conversation. It's about celebrating each other's differences of being able to acknowledge each other's differences."

What Is Equity?

Equity is supplying everyone within an organization the right tools, resources, and support they need to be successful in their role. Yvette Steele, an experienced DEI strategist, says equity is "is all about being able to create processes and systems that level the playing field so that everybody has an equal opportunity to achieve the same goals."

Equity ≠ Equality

Although equality is an important aspect, equity and equality are focused on different things and should not be confused. With equality, an organization is supplying the same tools, resources, and support to every employee, without regard for differences in how each employee approaches their work, or what they may individually need to be successful. Figure 2.1 illustrates the difference between equality and equity.

EQUALITY   EQUITY

Figure 2.1: Illustrating the difference between equality and equity

Three people of varying heights are trying to reach the fruit on the vine. If we gave boxes of the same height to each person *without* considering what their natural height is, then we may have only made it easier for the tallest person to get to the vine, while those who are shorter will still struggle to reach it.

Now, if we gave each person a box, keeping in mind their height, we would give them each a unique box. While each box may be different, we see now that each person can now reach the vine and grab the fruit.

Equity > Compensation

Ensuring that employees are paid a fair wage is an important part of creating equity within an organization. But this is only a fraction of that conversation. Access to advancement opportunities, rewards, and recognition are also very important factors to be considered. If employees do not feel that they are being paid fairly, or are being passed over for promotions, they are likely to become frustrated, disengaged, and ultimately leave an organization altogether.

What Is Inclusion?

Inclusion requires us to sincerely value and respect everyone, and the unique gifts that they bring to the table, as well as providing a safe space to be themselves. With inclusion, "everyone feels respected and there's a really high level of self-awareness . . . you can really sense that connection and that belonging," says King.

A favorite definition of inclusion comes from *The Inclusion Breakthrough: Unleashing the Real of Diversity (ReadHowYouWant, 2012)* :

> *"Inclusion is a sense of belonging: feeling respected, valued for who you are; feeling a level of supportive energy and commitment from others"*

Without inclusion, diverse talent may feel like they need to hide all or parts of their identity to be successful, or worse, prevent them from being harassed or fired from their positions. This is commonly referred to as *covering* and can take the form of an employee altering their appearance, avoiding behaviors that are widely associated with aspects of their identity, or avoiding contact with colleagues with similar identities. Because employees may go to great lengths to hide these parts of themselves, it takes a negative toll on their well-being, and ultimately their engagement with their work.

Inclusion, Explained Further

Figure 2.2 illustrates what DEI work strives for in terms of inclusion. On the top right and left we see both *exclusion* and *separation*, where diverse communities are completely left out of the primary group individually or as a group.

In the bottom left of Figure 2.2, we see *integration*. Here, a diverse group may be included in the primary group, but they are still "bubbled off" from everyone in the primary group. These diverse groups may be included, but only on the premise that they conform to the larger, primary group. People within these bubbles may feel like they do not belong, even if theoretically they are within the group.

With *inclusion*, on the bottom right, everyone is occupying the same space and with no boundaries between them. Everyone can come to their organizations as they are and are valued for it.

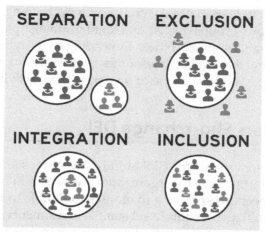

Figure 2.2: Comparing separation, exclusion, integration, and inclusion

Diversity Can't Thrive Without Equity and Inclusion

While it is commendable that organizations are devoting more time and resources to diversifying talent pipelines, focusing on diversity alone is counterproductive. Without fair and inclusive structures in place, organizations *cannot* keep or further develop a diverse workforce, or taken further, reap DEI's full benefits.

Per Steele, "you can get diversity, but without equity and inclusion, it's toxic. If we don't feel welcomed, or don't feel included or have an equal chance of achievement and success like everyone else and there's no equity . . . If you're going to live out the values or see the benefits of diversity, equity, and inclusion—innovation, greater creativity, greater problem solving, increased profits, all those things that come with having a diverse workforce—you won't if you don't leverage those three components."

Paraphrasing an analogy from an engineering manager, King offered, "Diversity is to have different players on the football team. Inclusion is to pass the ball to each player and equity is each individual having their own unique football boots."

What Is a Diversity, Equity, and Inclusion (DEI) Initiative?

A DEI program is considered a collection of activities, programs, and policies that encourage representation, participation, and fair treatment of everyone within an organization.

These can include specialized recruiting efforts, training, the creation of employee resource groups (ERGs), and forging strategic partnerships with outside organizations, among others. Depending on the type and size of the organization, and the number of resources at an organization's disposal, an organization may undertake one or several of these activities in tandem.

How Leaders Shortchange DEI

In a 2022 research report for Global Industry Analysts, the worldwide spend on DEI initiatives by organizations was $7.5 billion in 2020. By 2026, they expect this spending to more than double to $15.4 billion.

While DEI initiatives require a substantial investment of resources, history has shown that having big budgets does not translate into successful DEI initiatives. Because of the challenging nature of the work, many initiatives fail, even with the most carefully planned and well-intentioned efforts. Here, we'll examine the primary mistakes that organizations make that lead to the diminished results (or complete failure) of their initiatives.

For this discussion, we consider a DEI initiative to be failing when the organization yields little to no tangible or significantly measurable changes in their workforce demographics or culture in relation to the total effort they have spent.

Common DEI Pitfalls

Given how complex the topics diversity, equity, and inclusion are on their own, it should be no surprise to learn that implementing and executing a DEI strategy—even for those with vast resources—can be challenging and many things can go awry. Here, we discuss some of the more frequent reasons why some DEI strategies and initiatives flounder or fail.

No Overarching Strategy

Author and educator Lee Bolman once said, "A vision without a strategy remains an illusion." An organization's DEI efforts will likely be unsuccessful without:

- A solid strategy with tangible and realistic goals, backed up by the actions to be taken
- Knowing why the organization is taking these actions

- Knowing who will own carrying out those actions within the organization
- Measurements of success

No Commitment from Leadership

DEI initiatives must have strong, visible, and intentional support from senior leaders, especially those at the C-levels of their organizations. Topics, such as creating more pay and opportunity equity within the organization, requires management to lead the discussion and implement change, as they have the authority and ability to make those changes.

Employees are also more likely to buy into DEI initiatives when they see that their leaders are (sincerely) committed to them. "Leaders have to model the behaviors they want to see," says Steele. Without strong leadership support, these initiatives will not be successful.

DEI Work as a Checklist

Because the nature of DEI work will change and grow as an organization evolves, it should be viewed as an open-ended journey versus a checklist of tasks to be done. The checklist approach implies that if all the boxes, or "to-dos," are checked off, then all the work is done. Nothing more needs to happen. But as an organization grows and changes, the DEI strategy or work that was previously done may not well serve the newer version of the organization.

DEI work in organizations should be approached almost with an agile practices framework and mindset. Work is carried out and reviewed for progress, against performance metrics, with continuous feedback from all stakeholders, while carefully integrating new projects or tasks in the process. Work is ultimately performed in a loop, whereas a simple checklist does not allow for an opportunity to foster feedback, change, or improvement.

DEI Work as a Human Resources Function Only

While many of the activities in a DEI initiative are performed by human resources (HR) practitioners, DEI work should not be viewed as a pure HR function or in a siloed fashion. For DEI initiatives to be successful, efforts must be woven into an organization's policies and practices. Additionally, DEI efforts must be taken on by *everyone* in the organization versus a designated few. This includes cross-functional leadership.

Ignoring Intersectionality

Intersectionality is a concept where it is acknowledged that our unique backgrounds and experiences combine and overlap—they ultimately will shape how we experience the world. These dimensions of our identity can also influence the ways in which we experience discriminations and systems of inequality.

An organization's employees can be defined on several identity dimensions and trying to place employees in a single "box," so to speak, can undermine DEI efforts.

We (the authors of this book) are both straight, married, cisgender women who were born in the U.S. But we also are of different races, ages, educational and professional status, family composition, and socioeconomic status. Even though we were both born in the U.S., we grew up on opposite coasts, and thus grew up in completely different communities and cultures. Our gender identity alone does not fully define us, and assuming we both need support based on gender alone is shortsighted. What good employer support looks like for Bertina may differ vastly from what good employer support looks like for Susanne.

Not Establishing (The Right) Metrics

The success and effectiveness of DEI initiatives cannot be accurately assessed and evaluated without proper ways to measure in place. Admittedly, this is easier said than done. It's difficult to understand what metrics should be measured, how often they should be measured, and how to measure them. For example, statistics around who within leadership has taken part in DEI training can be helpful, but measuring leaders' proficiency in key DEI and HR skills is good to know, to help better identify where skills and knowledge gaps exist.

There's also the balancing of what success looks like for an organization versus how they compare to their peers. While it is helpful to know what industry peers are doing and to refer to their progress for general guidance, it is also important not to create DEI goals that are too ambitious, or not in line with the vision or needs of the organization. What is important is making steady, committed, and measurable progress.

DEI as a Marketing Campaign

Organizations with DEI initiatives that focus only on raising *awareness* of DEI issues and highly publicizing their diverse workforce, rather than focusing on actionable outcomes, risk having their initiatives viewed by others as inauthentic—perhaps a "marketing ploy" or as *performative allyship*.

With allyship, those in positions of privilege and advantage use that to amplify the voices of those who have been historically marginalized, and ultimately, bring about change. With performative allyship, messages of support are communicated and publicized, but no tangible action is taken. Those communicating their solidarity are looked at almost as heroes, but they themselves have taken *no* real action, and status quo remains.

Broadly advertising commitments and work in DEI is not an entirely bad thing. Many job seekers, particularly Millennials and Gen-Z, want to work for organizations that show a firm commitment to DEI. Where organizations run into trouble is when the positive pictures and communications belie larger problems, like having a low representation of women and people of color in non-administrative/operational professions and management, or when diverse talent expresses that they have no real sense of agency in their roles.

Ineffective Recruiting Practices

As discussed in Chapter 1, the continued reliance on recruiting talent from the same places, non-inclusive language on job descriptions, focusing on the wrong candidate criteria (e.g., irrelevant higher education requirements), and biased hiring practices will continue to yield a homogenous workforce.

Ineffective Diversity Training

Making DEI training that is rooted from a corporate risk mitigation and compliance standpoint mandatory can work *against* an organization's DEI efforts.

Harvard University sociology professor Frank Dobbin, and Tel Aviv University associate professor of sociology Alexandra Karev conducted a five-year study analyzing the effects of DEI initiatives for over 800 small to mid-sized companies. The study showed that mandatory DEI trainings saw no improvement in the representation of women or people of color in management roles. In fact, some populations saw the most employee attrition, with Black women leaving organizations at an average of slightly over 9 percent.

Steele offered more direct words: "If you have an organization where your DEI investments are more for the point of being compliant and to avoid litigation, then you're really not serious in the first place about [DEI]."

Diversity trainings that employ negative messaging, or "blame and shame" tactics, can also push people away. There is nothing wrong with directly acknowledging the fact that, historically, certain groups have enjoyed privileges because of their race, gender, social standing, and wealth. This has, at times, been at the expense of marginalized groups. However, trainings that frame these groups as outright villains who must atone for past transgressions and suggesting that they have not worked hard in their own right for their success only fuels their overall resistance to DEI efforts. Not to mention, it runs counter to what DEI work is trying to address which is to truly understand and value individual differences and contributions.

Ineffective Talent Development Programs

Mentorship and sponsorship programs can be a great way to identify and cultivate up-and-coming talent. Unfortunately, most of these formal programs lack clear guidance on what the goals of the program are: candidate acceptance criteria, and solid measurements to evaluate the overall effectiveness of the program.

Little time is spent in matching mentors to candidates, with those running the program basing matching decisions on skill alone versus their mutual availability, location, communication styles, conflict resolution approaches, and other criteria. If not paired with the right mentor or sponsor, participants can come away feeling like the programs did not meet their needs and expectations (more on this in Chapter 5).

Examining program participant demographics is also important. If there is a lack of diverse participants, it can lead to more questions— how accessible is the program to diverse audiences? Are the participant requirements too restrictive? How is the program's existence being communicated and socialized?

No Accountability

Organizations that do not implement effective means to hold their employees accountable for their work and their commitments risk negative consequences. Besides wasted efforts and the potential for bad actors to take advantage since there's no appropriate retribution, lack of accountability can erode trust.

The same is true for DEI initiatives. If there's "no skin in the game"—meaning there are no personal risks or consequences for when efforts are not executed properly and subsequently fail—then everyone loses in the end. Holding leaders to their DEI commitments is necessary for success and sustainability.

It is also that accountability is *distributed*, meaning accountability for the success of DEI initiatives is at all levels, not just with senior leadership. Every employee, regardless of business function, must be held responsible for carrying out the organization's DEI vision and mission.

No or Low Compensation and Recognition for DEI Work

Often, many employees who take on DEI work within their organizations are doing so in addition to their primary job responsibilities, and for no additional compensation. This is time that could be spent pursing professional development opportunities or personal downtime. This work also seems to be disproportionately taken on by women and people of color.

Make no mistake, while DEI work can be rewarding and fulfilling, it is *serious and challenging work*. DEI work can require having tough, uncomfortable conversations, challenging systems, and old ways of thinking. The work can take an emotional and mental toll, on top of physical effort, and lead to burnout. In 2022, LinkedIn reported that the average tenure of someone who holds the role of chief diversity officer or similar is only 3 years.

While it is not always possible to financially reward employees for their efforts (although in an ideal world, it should), tying DEI work to recognition and advancement opportunities goes a long way.

Not Listening to Employees

While leadership buy-in is important, taking the time to listen to employees and hearing their honest and candid feedback is also important. Hearing employee feedback and concerns can help leadership validate if their efforts are having the desired outcomes or inform them of where their efforts are falling short.

Whether it's through employee surveys, town halls, or even having one-on-one conversations with employees, these offer opportunities to receive valuable insights. DEI initiatives without a means for continual dialogue between employees and leadership cannot hope to improve or grow.

No Transparency

Employees expect a level of transparency with understanding the overall health of their organizations. While it is not always a hard requirement for organizations, employers who regularly communicate to employees on matters like the financial health of the business and compensation strategies and practices are likely to have higher employee satisfaction and engagement than those who do not.

As DEI is (or should be treated as) an integral part of an organization's strategy, regularly communicating an initiative's progress and key metrics to employees is also important. Employees want to know that their organization's commitments to DEI are not just surface level and a "nice to have." Rather, it is an important element for the health of the organization, and they are doing the hard work necessary to make progress.

"Copying and Pasting" DEI Initiatives from Other Organizations

The DEI work that takes place in a multinational high-tech company will—and should—look vastly different from the DEI work that is undertaken by a small tech startup. This is not just because of the size of the company, the industry that it is in, or the resources that it has at its disposal. Rather, DEI work needs to address the needs of an *individual* organization.

There are certainly lessons to be imparted and taken from the success stories (and even failures) of others. But taking cues from other organizations' DEI without taking into account an organization's current state and communities, and adjusting those efforts as necessary, will lead the initiative to failure, quicker.

The case studies that are presented in *Innovating for Diversity* provide opportunities for further reflection and consideration, but should not be copied and implemented verbatim.

Going Alone

We don't know what we don't know. Even for the most experienced DEI practitioners, there are blind spots or gaps in their knowledge of current DEI best practices.

And that's honestly OK. No rational person would expect an organization to know everything about starting and implementing a DEI initiative or executing it flawlessly. If an organization knows that this gap exists internally, it is best to enlist the help of others, when possible, versus creating an initiative alone. Not only does this minimize the amount of wasted time and effort, but it can also help minimize unintentional messaging, gaffes, and harm to employees.

Another reason to seek additional help is that we may not be the best assessors of our blind spots and knowledge gaps. Having an independent and objective outside party can help organizations better understand where gaps and biases exist and suggest prescriptive strategies to address them.

The Consequences of Ineffective DEI Initiatives

The effects of running ineffective DEI initiatives go far beyond wasted time and financial resources that don't yield DEI's benefits.

There's the risk of losing top talent. "People who know their value are not going to stay within your organization, then you're going to have a huge retention problem," says Steele. "And you're going to be constantly where the Great Resignation will probably hit you harder than anyone else who's working hard to create an environment and an ecosystem so that people will want to stay."

King takes this a step further and offers that the hit can extend to an organization's reputation. "I think that it can really hinder, potentially, [an organization's] reputation as a business when it comes to hiring new employees and staff. Markets can be very small. People can talk, word can get around. And if you're not acknowledging the true value of a diverse, equitable, and inclusive environment and culture, and let's say that you do have a minority in your business, and they really feel the brunt of that . . . they feel that pain—social media now more than ever is rife. People can slam some companies and be quite abrasive in the way they do it and not hold back. And you don't want that as a business, you don't really want to be tied to that. All it takes is one person to say one bad thing, and then it's so hard to get back from that."

Conclusion

In this chapter, we reviewed the definitions of diversity, equity, and inclusion in greater detail, and provided visibility on why DEI efforts can be so challenging to create and implement.

It's important to remember that even with these cautions and challenges, it is possible to create and implement effective DEI initiatives. In the next few chapters, we'll talk about conditions for success and explore how basic principles of innovation can inspire DEI approaches that address underlying barriers and improve long-term outcomes.

Summary

- Diversity is any dimension in which we can differentiate distinct groups from one another.
- Equity is providing individuals with the tools and resources they specifically need to be successful in their roles.
- Inclusion is the state where all employees in an organization feel valued, respected, and safe to be themselves.
- A diversity, equity, and inclusion initiative (DEI) is a collection of activities, programming, and policies to ensure the representation and equitable treatment of diverse talent.
- Despite large investments, DEI initiatives fail for a variety of reasons, including lack of leadership buy-in, approaching DEI as a "one and done" activity, not attaching metrics and goals, and not building in accountability.
- Beyond the loss of money and time, continuing to run ineffective DEI programs risks raising employee attrition and loss of competitive advantage.

CHAPTER 3

The Virtuous Cycle of Innovation and Diversity

Innovation has long been a central driver of business growth; without it, entire industries become extinct. While there are many factors that contribute to innovation success, a diverse and inclusive culture that is open to change is paramount. In recent years the linkage between diversity and successful innovation has been well documented. However, the opposite relationship is less explored—how innovation can drive success in advancing diversity. The diversity, equity, and inclusion industry is a big business and more often focused on compliance rather than confronting the deeper, persistent challenges brought on by the Fixed Practices and Fixed Attitudes discussed in Chapter 1.

This chapter outlines a general innovation framework we will reference throughout the book in discussing case studies of companies and organizations creatively solving for diversity.

Key Concept: Diversity leads to greater innovation; and innovation drives improved diversity.

The Power of Innovation

Throughout business history in America and around the globe, innovation has been the catalyst for wholly new industries and profound wealth creation. In recent decades, brilliant, well executed ideas have transformed how we communicate, travel, invest, heal, learn, and work. Few aspects of everyday life are untouched by the accelerating pace of innovation fueled by the availability of venture capital (VC), private equity, sovereign wealth funds, government funding, and corporate commitment to research and development (R&D) with the expectation of financial returns. And, of course, innovation is fueled by a boundless consumer appetite for the new and novel. More altruistically, innovation is driven by a human instinct to increase living standards and create a safer, healthier environment. It was, after all, through rapid innovation delivered through a collaboration of talented scientists and healthcare professionals and concentrated capital infusions that brought COVID-19 vaccines to market in record time. Innovation drives not only corporate profits and competitiveness but is essential to solving the most vexing problems of our time, and those we anticipate will confront future generations.

Economists feared that investment in innovation would lag during the height of the COVID-19 pandemic as the global economy slowed and unemployment rates jumped to levels not seen since the 1930s. Yet scientific output, R&D expenditures, intellectual property filings, and VC funding grew in 2020, building on a peak of prepandemic innovation investment in 2019, according to the 2021 Global Innovation Index developed by the World Intellectual Property Organization.

While innovation expenditures dropped in industries hardest hit by the pandemic—travel and transportation—innovation investment surged in sectors that supported recovery: software and information and communication technology (ICT), ICT hardware and electrical equipment, and pharmaceuticals and biotechnology. Companies with two- to three-year implementation plans for widescale digital transformation—integrating technology solutions to all parts of their operations—dramatically accelerated implementation in a matter of months, in part so that employees could effectively operate remotely and to improve supply chain management. A 2021 survey of 1,200 C-suite executives by Citrix reports that 80 percent of leaders believe their companies have entered a new phase of hyper-innovation driven in part by a tech-enabled hybrid workforce.

Companies that excel at bringing innovation to market recognize that financial commitment to R&D and technology is but one part of their success. They also have a deep understanding of their markets, align innovation strategy with business strategy, and establish strong project management systems that identify the best new ideas for development. They also benefit from visionary leaders connected to the innovation pipeline and a culture that embraces diversity, change, and growth. Companies with these characteristics are highly efficient with their R&D investments and more likely to be what the consulting company PwC calls "high-leverage innovators." In PwC's 15-year study (2002–2017) of 1,000 publicly held companies, high-leverage innovators outpaced their industry peers in seven key indicators of financial performance, including operating income growth, sales growth, and market capitalization growth. Remarkably, these companies on average dramatically outperformed their competition during the 2007–2012 business downturn and the early years of recovery, suggesting that smart innovators are better positioned to thrive during economic contractions.

Why Companies Get Stuck

With so much evidence supporting the necessity of innovation, why don't more companies do it well? It turns out the answer is remarkably similar across industries and sectors. Let's look at the three themes that consistently emerged during our conversations with dozens of successful tech entrepreneurs, seasoned executives at Fortune 500 companies, research scientists at top labs, and leaders at nonprofit and government agencies.

Innovation Is Simply Not Prioritized

An organization may represent the importance of innovation strategy, but not genuinely support it with resources or leadership, instead preferencing the demands of the current business model. The core business may be *successful enough*, buttressed by incremental improvements over time that provide just enough gloss to lull staff and leaders into believing real innovation is underway. While incrementalism can be a smart service or product development strategy, it can sink a company when a true disrupter emerges as a competitor. One leader we interviewed at a software development company reflected on how a singular focus on marketing an existing product to new customers blinded the management team to the need for product innovations to maintain their installed base of customers. Building a new audience seemed like a relatively quick and lucrative win. The current set of customers were taken for granted, seemingly satisfied with periodic product upgrades. When a set of upstart competitors made headway with their core customer base, the company stalled. For too long they had underinvested in transformative innovation; they had relaxed their processes to identify and support good ideas, lost their discipline for long-term planning, neglected to listen deeply to the needs of their best customers, and failed to reward the strong innovators on their team. Many of the very leaders who could have reignited the culture and shifted priorities left the company to found their own start-ups.

Inertia Is the Mortal Enemy of Original Thinking

We celebrate the idea that "success breeds success," but it can also lead to complacency and launch a set of processes and procedures more intent on protecting that success for as long as possible, rather than ensuring its evolution or identifying the next big idea. We were struck by the number of leaders we interviewed who reflected on their observations about a "culture of coasting" after a new product was successfully launched or during long stretches of business stability. With the risks inherent in innovation, why change something that's working? This question becomes even more challenging to tackle at the individual leadership level when security and personal wealth like bonuses and stock options are at stake, making inertia a quiet but potentially lethal enemy.

During our interviews, we also heard stories about numerous formal systems designed to operationalize best practices and create standards built around the delivery and implementation of a successful line of business. Now, there is no question these systems are a vital management practice, especially as employees move on and institutional memory dims. The danger is when a playbook or checklist so deeply ingrains a set of activities that an organization comes to accept only one right way of doing things. One leader we interviewed described the hazard as "managing on autopilot." We may have checked the boxes but failed to realize a whole new set of boxes were needed to respond to a changing environment.

The Power of Humility Is Overlooked and Undervalued

To think you can solve a large-scale problem that has long vexed others or design a wildly successful product no one knew they needed requires enormous confidence and a healthy dose of moxie. But when these characteristics grow unchecked, they can morph into arrogance, a dangerous toxin to innovation. The barrier quality that keeps confidence in productive territory is simple humility. Leaders we spoke to shared numerous examples of how a lack of humility produced cultures resistant to collaboration, curiosity, and change.

An executive at a health diagnostics start-up reflected on a time when he worked for a division of a large healthcare company established solely to innovate new digital health products. The head of this innovation unit inspired confidence and recruited some of the most accomplished researchers in the field, as well as bright technologists without deep experience in healthcare. The executive we interviewed described the team as a "mash-up of great minds from across different disciplines," which created an atmosphere of unlimited possibilities. Candidates for roles in the division were screened for their ability to work constructively on a team, check their ego, and even their willingness to mentor others. Within two short years, there was a promising pipeline of products, many of which had the potential to compete in market segments additive to the company. When the founding leader left the company, a new boss arrived with deep experience in the established core business and an unyielding set of convictions about the best new product bets. While his decisiveness may have played well with the CEO, his lack of inquiry and interest in deeply understanding the robustness of the new product pipeline signaled a change in culture and approach at odds with the creative, curious, and humble hive that existed under his predecessor. Many of the most highly regarded researchers left to pioneer new products elsewhere, leaving behind gaps in the internal innovation pipeline.

Another leader we spoke with discussed the importance of collaboration, particularly with those whose expertise is complementary, in the innovation process. But he cautioned that collaboration only works when approached with humility, that is, a real acknowledgment that others may have a piece of the puzzle we haven't been able to find ourselves. Humility gives us permission to not have all the answers, to fearlessly ask penetrating questions, and to learn deeply from our own failures and from the success of others. Without humility, it is nearly impossible to get at the core truths and insights that lead to breakthrough innovation. Without it, we can become too secure in our worldview and our confidence can turn to arrogance, further suppressing the initiative to innovate. Even if innovation is purchased through acquisition, a healthy dose of humility goes a long way in ensuring successful integration.

Without fail, each leader we interviewed referenced the outsized importance of culture in creating conditions for successful innovation, including the cultural characteristics of courage, risk-taking, trust, collaboration, and leadership, which we will explore in further detail later in this chapter.

Diversity Drives Innovation

The role of culture in fostering innovation is well-documented and reinforced in our own interviews, but the specific contributions of diverse and inclusive teams to the success of new products and business lines have gained significant attention from researchers in recent years.

With corporations increasingly operating in a global environment and responding to a growing minority and multiracial U.S. market, attracting and engaging diverse talent should be viewed as a core business strategy, not just a program or initiative. A 2021 Forbes survey of more than 300 senior executives of multinational firms revealed an overwhelming belief in the linkage between diversity and successful innovation. Over 48 percent of all survey respondents strongly agreed that diversity is crucial to driving innovation with the number increasing to 56 percent among respondents from the largest companies with over $10 billion in annual revenue. More than three quarters of respondents indicated their companies will give more focus to leveraging diversity for innovation and other business goals in the next three years.

These executive responses are driven by more than perceptions and compelling anecdotes. Companies that have prioritized diverse leadership teams are more innovative and sustain better overall financial performance. A 2018 study by Boston Consulting Group (BCG) found that companies with above-average diversity among their leadership ranks have a greater financial return on innovation and higher earnings before interest and taxes (EBIT) margins. In their research of more than 1,700 companies in eight countries, BCG found that those with above-average diversity on their management teams reported revenue from products and services launched within the past three years—"innovation revenue"—was 19 percentage points higher than companies with less diverse management teams (Figure 3.1). Companies with better diversity scores also reported better overall financial performance with EBIT margins that were 9 percentage points higher.

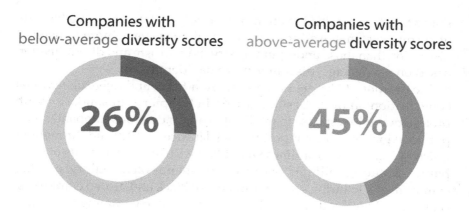

Companies with
below-average diversity scores

26%

Companies with
above-average diversity scores

45%

Figure 3.1: Comparison of innovation revenue between less vs. more diverse companies

Rocío Lorenzo et al., 2017/Boston Consulting Group.

With evidence supporting the linkages between diversity, innovation, and financial performance, why do U.S. companies still struggle to achieve full representation, especially racial representation, in their professional and executive ranks? As noted in Chapter 1, Black, Hispanic, and Asian professionals continue to trail their white counterparts in assuming executive roles and double-digit wage differentials persist for minority and women workers compared to white male workers in similar roles. As we reviewed in Chapter 2, traditional approaches to building more diverse teams can fall short, making real progress frustratingly slow. Arguably, formulaic approaches aren't addressing root problems or changing the underlying systems anchored in Fixed Practices and Fixed Attitudes that are working against diverse populations.

Let's turn the truism that diverse teams can yield better innovation upside down. It's time to apply better innovation to building more effective diversity practices.

Think about that for a moment. The greatest innovations disrupt long-standing systems and challenge our worldview about what is possible. The success of electric vehicles, led by Tesla, has reshaped our energy infrastructure, with the promise of making charging stations as much a part of our physical landscape as traditional gas stations. The invention of the blockchain is not only driving a movement of decentralized finance that is disrupting the banking sector, but it will change how we vote, secure supply chains, and manage our most sensitive documents.

Powerful ideas that lead to an improved future state transcend barriers of current systems. And it's hard to imagine systems more in need of disruption than those underpinning structural racism and discrimination based on gender, age, disability, or sexual identity.

The murder of George Floyd sparked a frank and overdue national conversation on race. As summarized in Chapter 1, in solidarity with protesters and in response to public upheaval, numerous companies took public actions to show support for racial justice. In 2020 and throughout much of 2021, news headlines were filled with announcements of companies donating millions to causes advancing racial equity, making commitments to hiring greater numbers of Black and brown employees, celebrating Juneteenth, and taking concrete steps to becoming anti-racist.

The myriad actions undertaken by hundreds of U.S. and global companies should be celebrated as steps toward systems change, and in some cases as examples of real innovation. But the problem is far from solved. In 2020 McKinsey & Co. conducted a study of perceptions of diversity and inclusion initiatives among dozens of companies across three sectors: financial services, technology, and healthcare. They used research techniques such as "social listening" on U.S.-based online platforms, designed to collect employee reflections on their employers' diversity and inclusion progress that were likely more candid than comments offered in internal employee-satisfaction surveys. The researchers learned that employee sentiment on diversity was overall 52 percent positive and 31 percent negative. However, sentiment on inclusion was substantially worse at 61 percent negative.

Even with so many ambitious CEO proclamations and sweeping headlines, DEI efforts at many organizations remain largely limited to training and workshops, employee resource groups (ERGs), and opening new employee recruitment channels, all of which are vital components of an overall DEI strategy. However, a handful of stand-alone initiatives that change little year over year can signal a company cares more about checking "DEI activity" boxes than solving underlying problems and owning accountability. Hiring a chief diversity officer supported with meaningful resources, adequately supporting the Human Resource division, or retaining a smart DEI consultant can set a course of real change, but only if senior leadership, including the Board, is prepared to set goals, measure results, demand accountability, and course-correct with the same focus as any other part of their business.

Remember *"inertia is the mortal enemy of original thinking"* from earlier in the chapter? Resourcing DEI activities that have become rote leads to the false comfort of progress. The field of DEI has itself become a big business, making it easier than ever to implement activities that feel good and allow executives to point to actions, but may not alone produce needed outcomes. The global business of DEI, including workshops, ERGs, and corporate training is expected to reach $15.4 billion by 2026. DEI is a juggernaut of an industry filled with consultants advocating for multiphase processes and programs that at best support real outcomes and at worst signal to employees that DEI is simply a passing trend or compliance mechanism, not necessarily a deep commitment driven by the CEO and broadly shared by leadership at all levels.

The very practice of DEI is threatened by the same three limiters to innovation noted earlier: lack of prioritization, inertia, and arrogance.

But what if we applied the same rigorous frameworks we use for product and technology innovation to the practices of sourcing, retaining, developing, and advancing diverse talent? Taking that a step further, what if we encouraged *innovation for diversity* to occur beyond the boundaries of the HR department and celebrated the winning efforts of individuals and teams across an organization, working in collaboration with HR professionals? Bringing a world-changing idea to market is rarely the work of a singular genius or, for that matter, the department of the chief innovation officer. Similarly, innovating new ways to build diverse and inclusive workplaces cannot be solely owned by the HR leader or even the chief diversity officer.

Innovation Principles

Let's first take a deeper look at the behaviors and organizational char- acteristics that create conditions for innovation, regardless of industry or function. Innovation broadly requires a nuanced mix of values and collective strengths that defines a culture where innovation can flourish. It also requires mechanisms to keep the threats to innovation discussed previously—lack of prioritization, inertia, and arrogance—in check. Thinking back to our executive interviews, the five cultural character- istics necessary for innovation most commonly named include courage, risk-taking, trust, collaboration, and leadership.

In the following chapters we will discuss a series of case studies of companies and organizations creatively solving for diversity and explore how these five characteristics advantaged breakthroughs in DEI practices. We'll also examine how successful companies avoided the pitfalls of allowing diversity to fall down the priority list, falling victim to inertia, or stumbling for lack of humility.

Courage

To be courageous is to show strength in the face of pain or take an action that is frightening or unpleasant. We often associate courage in its greatest form with inspiring leaders and fearless inventors; people who sacrificed personal safety to achieve social change or create world-altering products and technologies. But courage can also be defined by the small yet meaningful actions in the workplace that are necessary for breakthroughs. How many times have you been grateful to the bold colleague who spoke up in a meeting and expressed the concerns and questions no doubt on the minds of others who were afraid to speak out? Can you think of a time when you've had the audacity to recommend an alternative to a way of work that is so deeply ingrained that it is widely accepted to be the right and only way?

Michele Garfinkel, a researcher and science policy expert, spent her early career studying innovation theory while at the Center for Science Policy and Outcomes. She defines professional courage in the context of innovation as the *opposite of conservatism*. Her definition is founded on experiences working on high-profile scientific breakthroughs as well as the foundational activities of collecting and analyzing data. In our interview she reflected on the role of courage in the culture at the J. Craig Venter Institute that led to the creation of the first synthetic genome as one that blended fearlessness with humility. Founder Craig Venter modeled courage in scientific exploration and asked his staff to be fearless in their work. Craig's challenge to be fearless was actually permission to be courageous in scientific inquiry and came with his commitment of back-up and support. But courage was needed for more than the research. In her role working in the policy division at the Institute, Michele was struck by the humility of lead scientists who were "brave enough to hear the critiques of their research from the policy, ethics, and regulatory perspectives and to acknowledge they didn't have all the answers, just because their science was groundbreaking." Even in the early days of their synthetic genomics experiments, the group had the courage to engage bioethicists, legislators, non-governmental groups, and the general public to examine and challenge the societal and ethical implications of their work. The team was well prepared to respond to the Presidential Commission for the Study of Bioethical Issues tasked with investigating the implications of synthetic biology. The Commission's final report recommended a balanced approach of White House-level oversight of synthetic biology research and stopped short of recommending new laws or changes to existing regulations that could have interfered with subsequent innovations leading to new treatments for diabetes and leukemia, among numerous other achievements.

Michele's experience at the J. Craig Venter Institute demonstrates courage on a scale that impacts the future of medicine, food, energy, and even the clothes we wear. But we were also struck by her reflections on the value of courage in the everyday tasks of good science—data and analysis. Courage may look like questioning the integrity of a data set days before a grant report is due or recommending an analysis departing from convention. In practice, these smaller acts of courage take enormous energy and personal resolve.

Others we interviewed reinforced that courage not only requires confidence and fearlessness, but a tremendous reserve of personal resilience and stamina. When asked to think of actions demonstrating the underlying courage needed for innovation, executives responded with:

- Courage to demand excellence and not follow someone else's playbook
- Courage to speak up, challenge norms
- Courage to examine the real underlying problem we're trying to solve
- Courage to be the first
- Courage to be invested in real change, not quick fixes
- Courage to confront resistance to change in an organization
- Courage to be an owner
- Courage to not just go with the flow
- Courage to not just help people navigate systems that were not built for and by them, but to change those systems
- Courage to look beyond boundaries
- Courage to press the pause button
- Courage to be uncomfortable

Courage is a highly intentional act that requires time, thought, and energy. It is the personal quality required to productively confront risks that cannot always be quantitatively measured and managed. Without it, the best we can hope for is moderate incrementalism, but not true innovation.

The path of least resistance rarely leads to breakthroughs.

Risk-Taking

It is hard to imagine innovation without a healthy measure of risk-taking. By definition, innovation represents something new and requires a departure from what is safe and known, which for many of us is uncomfortable territory. The prospect of failure can be paralyzing, keeping us bound to a routine that gets work done with little challenge or disruption.

Yet risk in evaluating just about any innovation can be identified, quantified, and managed. Risk mitigation strategies that define accountabilities and consequences can then be implemented. So, what's so hard? Why does risk-taking emerge as such a huge barrier to innovation if it can be carefully managed?

Jesse Hillman, the chief information officer at Tessco, knows about risk. Jesse served in the U.S. Navy as Special Operations Officer. While the types of risk most of us confront in the workplace rarely come with life-or-death consequences, risk that cannot be neatly managed does require a certain amount of bravery. In talking to Jesse about his own approach to risk-taking throughout his corporate career, he describes two requirements: a clear-eyed analysis and an internal dive into one's own personal appetite for risk. The first he dispatches as a baseline requirement for any competent manager—the ability to do scenario planning, assign probabilities to potential outcomes, and quantify the impact of those outcomes on the health of the enterprise. In Jesse's experience, it's the second, more personal relationship to risk that is the bigger barrier to innovation.

Jesse recalled the time he was tapped to lead a significant portion of building a new enterprise resource planning (ERP) system for a global company. The time and financial investment were material, as was the risk of failure. In fact, similar efforts to update the company's ERP system had stumbled twice before under different leadership. Before taking on the assignment, Jesse thought carefully about the risks he believed he could manage, as well as his personal appetite for professional risk. On the line was his reputation, possibly his job, embarrassment, and failure to fully leverage the new ERP system as an opportunity to innovate much-needed processes at the company. In thinking about his decision to pursue the assignment, Jesse talked about his "confidence to be accountable" as a driver. In fact, he made "personal accountability" for each phase of the ERP development and implementation a requirement for each of his team members. Those on the team with a low threshold for accountability (and in Jesse's estimation, a high degree of risk-aversion) voluntarily left the company or were re-assigned to other roles. Jesse had the opportunity to build a team with the courage, skills, and resilience to own the project. They operated within a culture that supported failing fast, learning and improving constantly, and persisting. The project was one of the most successful enterprise-wide systems changes in decades, and yielded not only substantial cost savings, but also a platform for improved communications and decision-making.

With the pace of work accelerating and, for many remote and hybrid employees especially, the lines between work and personal life blurring, taking time out of the routine to assess risk requires planning and intentionality. Even those temperamentally drawn to well-considered risk-taking can be beat down by endless to-do lists.

Anything worth changing for the better is worth taking a risk for.

Trust

There are few values more highly correlated to strong team performance and empowerment than trust, or more negatively correlated to high levels of burnout and disengagement. Trust is predicated on mutual respect, shared values, and a belief that colleagues and leaders will behave in a manner that benefits the goals of an organization rather than for personal gain. In her landmark research on psychological safety within teams, Harvard Professor Amy Edmondson revealed how trust supports an environment where team members are more open to seeking critical feedback, asking questions, embracing failure without embarrassment, and exhibiting learning behaviors. The values of trust and psychological safety are especially critical during periods of high uncertainty when rapid information sharing and tolerance for challenge are necessary for navigating a new path. The act of innovating is itself a highly uncertain endeavor that demands exploration of the unknown and unproven.

In the tentative months following the COVID-19 outbreak, leaders struggled with major decisions impacting the health of their workers, service to their clients, and overall business viability. Absent the clarity hard data or national mandates can provide in a time of crisis, we witnessed successful leaders draw upon the reserves of trust built over time with their teams and external partners to innovate new approaches that helped them thrive during uncertainty. And, for some, these pandemic-induced innovations permanently changed their business model.

In adult education, a sector not widely known for disruptive innovation, we saw quick pivots from in-person to online instruction and career services with little drop-off in student retention or performance.

At NPower, we shut down our in-person training operation in March 2020 for two weeks. In that short period, we redesigned the delivery of our curriculum, overhauled our pedagogy, and re-invented our work-based learning model since we anticipated many companies would eliminate their internship programs. For the first cohort of the pandemic, over 400 students, we had one of our highest graduation rates in five years. We experienced innovation in hyperdrive—even while working remotely—in large part because we trusted our colleagues to make the best decisions to support our students, who are our clients.

That understanding of trust didn't emerge during the throes of crisis but was developed over years of trial and error and program experimentation, especially within NPower's Instructors Institute. The head of the Institute, Robert Vaughn, intentionally created an environment long before the pandemic where his team members felt comfortable openly sharing concerns and new ideas. They sat in on each other's classrooms and offered unfiltered feedback to help a colleague improve a lesson plan or lecture.

When a new training module was introduced, instructors readily compared experiences with their peers to assess their own efficacy in teaching the new material. Robert also encouraged a healthy sense of humor, which helped even the most reserved team members become more open and more fearless in asking probing questions to get to the best answer. And above all, this team shared an exceptionally strong commitment to ensuring all NPower graduates left the program with better opportunities for economic success. The high level of trust, together with the mutual respect and camaraderie built over time, fueled an impressive response in the wake of COVID and has had a permanent impact on NPower's training model, which is now more agile and responsive to student and employer feedback.

But paradoxically, innovation can falter when there is *too much* trust.

Trust flourishes when conditions of reliability, integrity, and mutual caring exists. A study by Francis Bidault, Professor Emeritus at the European School of Management and Technology, and Alessio Castello, Professor of Innovation at the International University of Monaco found that the qualities of integrity and reliability support creative problem solving and innovation. Mutual caring, on the other hand, can lead to excessive accommodation and a reluctance to challenge colleagues on their ideas and actions. For many of us, working with colleagues we genuinely *like* and care about is motivating. We will travel the extra mile to help a colleague with whom we have a personal connection and know the favor will be returned. Friendly bonds no doubt make for a more joyful and even more productive workplace, especially during times of crisis.

The caution is to prevent affection, and the fear of damaging a friendship, from leading to complacency and blind trust such that colleagues no longer question ways of work. Bidault and Castello's research shows that innovation increases with trust until it hits a "sweet spot" after which innovation decreases as trust becomes greater, as illustrated in Figure 3.2.

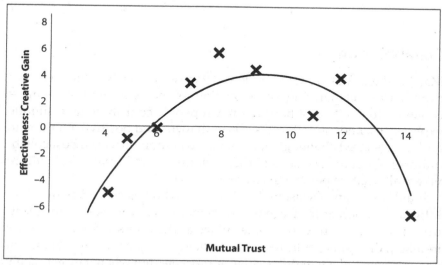

Figure 3.2: Effectiveness of the partnership in relation to the level of mutual trust
Entreprendre & Innover-De Boeck Supérieur, 2019

An operations team at one technology company we know well was especially high performing in reaching daily and monthly production results. The ten-member team had worked together for many years, socialized with one another outside of work, and was led by a charismatic and devoted leader. Others within the company admired and even envied their esprit de corps, low drama, and high trust. Within the boundaries of their routine deliverables, they outperformed every other team in the organization tasked with the same work. Part of their success was driven by their collective ability to innovate incremental improvements to current processes. However, when a modification to product design required innovating entirely new processes, the team floundered. They struggled with constructively challenging one another and questioning the quality of each other's proposed solutions. When a leader from another part of the organization stepped in to facilitate a series of problem-solving workshops, the team felt they had permission to be more critical—not of each other as individuals, but of each other's ideas—and ultimately landed on an approach that was adopted elsewhere in the company.

Trust is a good thing, but like with many good things, too much can be harmful.

Collaboration

Many of the executives we interviewed were eager to talk about their own experiences collaborating for innovation. We heard a common theme—collaboration at its best happens when people contribute their unique experiences, knowledge, and skills with focus and positive intention to address a shared challenge or achieve a common goal. The best collaborators respect the opinions of their teammates and create a subculture where alternative perspectives are encouraged.

Frank Baitman, a seasoned tech leader who has served as the chief information officer for large federal agencies as well as scrappy start-ups observed, "Innovation takes place at the intersections—whether industry, discipline, cultures, or technology. People who are steeped in one area won't likely see alternative approaches, but when they genuinely connect and collaborate with people who have different backgrounds, new and better ways to tackle a problem are uncovered. I've seen this in science, business, and public policy."

Katty Coulson, vice president of information technology for Oracle NetSuite, elaborated on this idea by noting, "Complementary skills on any team powers the engine of innovation. Variety of thought and experience is necessary since it's rare for any one team member to be expert in all the skills required to solve a problem." Kathy acknowledged that the act of collaboration has been at the center of virtually every successful innovation she has been a part of throughout her career. She observed, "collaborations are really partnerships, formed through the bonds of relationships built on trust." In her digital transformation work at Oracle, she depends on partnerships with both her internal clients and peers to get to the heart of the business challenges she and her team address through improved technology. "You can't just pass requirements over the fence, but rather you must collaborate to deeply understand the goals and priorities of the internal client." It takes time, patience, and investment in relationship-building to produce the innovations that have driven both performance and scalability in Kathy's client group.

Craig Cuffie, chief procurement officer at HSBC, expanded on both Frank Baitman's idea that innovation happens at intersections and Katty Coulson's belief that collaborators must also be partners. Craig reflected on his prior role as Vice President of Global Supply Chain at Intuit where customer success is a core value. The focus on customer input is what drives much of the company's product innovations. During his tenure there, he flipped the idea of customer-driven innovation to encourage supplier-driven innovation. With Intuit in the role of Customer, he organized a regular "Supplier Summit" to challenge the company's largest vendors to share ideas for how Intuit could improve. He observed that the best suppliers are "students of our business" and often present creative approaches for how the company can innovate on both product design and business process. This is a great example of boundary-blending that invites innovation at the intersection of customer and supplier.

Echoing throughout all of our interviews was the importance of leaders to see and value the unique skills and experiences of team members. Moreover, leaders who consistently innovate are adept at curating teams with a mix of expertise and fostering a culture of mutual respect, accountability to one another, and trust.

> *Collaborating for innovation requires partnerships across skills and expertise, built on trust and mutual respect.*

Leadership

Leaders play an outsized role in creating conditions where innovation thrives. Without a culture that supports and rewards creative thinking and problem-solving, organizations stagnate and ultimately fail. This doesn't mean a leader needs to personally be a genius innovator, but a leader does need to consistently demonstrate actions and behaviors that build a culture where others can be. This is true for leaders at every level of an organization.

When Marcus Valentine began a new assignment to lead product operations and strategy at a California-based software company, he inherited a team filled with individuals he describes as smart and driven, but who didn't quite fit in at other departments. Prior to Marcus's arrival, his department had become the place where those whose way of thinking or working didn't conform to cultural expectations elsewhere in the organization landed. Marcus saw vast potential in each of his new team members and surmised their lack of success was connected to cultural norms that suppressed their real strengths and discouraged their ideas. So, he set about creating a different culture. The reward would be a high-performing team that delivered on the CEO's expectations for process improvement. The risk would be personal failure, possible termination for the entire team, and loss of valuable time and resources for a high-growth company that couldn't afford distractions.

Prior to joining his new position, Marcus worked in private equity, product development, and in chief of staff roles for two CEOs. He had a set of experiences that honed his ability to view strategy from 30,000 feet but he also felt comfortable deep in the weeds. As he got to know his new team and department, he quickly realized a whole set of recurring business analytics—program dashboards, product status reports, competitive intelligence, price tracking—were either missing across the organization or lacked clear ownership. He saw a problem no one else did. Both the CEO and Chief Product Officer valued his observation and agreed he was the right leader to tackle the challenge.

The first thing you notice when you meet Marcus is the seeming contradiction of his blazing intelligence and easy manner. You think there is not a corner he can't see around, or a problem he can't solve. He's also the kind of person you wouldn't mind being stuck in an airport with. Marcus possesses a set of leadership characteristics—vulnerability, compassion, empathy, patience, directness, clarity—that was needed to build a culture supportive of what he describes as his ragtag team in their new mission to dramatically innovate business intelligence systems so other divisions had the right analytics to make the best decisions. This challenge became more urgent as the company acquired several smaller firms, each with their own systems that had to be integrated.

In his approach to culture-building, he led with vulnerability, curiosity, and empathy. Marcus drew upon his own experiences as a Black gay leader, reflecting on times in past roles when he felt constantly judged or discouraged from speaking openly about his personal life. The lack of safety and trust hindered his ability to be fully engaged and produce his best results, which is a crippling outcome for someone with Marcus's intellectual firepower. Knowing how a sense of insecurity impacted his own performance, he consciously modeled a different set of behaviors with his new team that included careful, patient listening of what each of his team members needed to succeed, and freely shared what he believed were his own shortcomings. He asked countless questions for deeper understanding and encouraged his staff to do the same with one another. When team members recommended novel technology solutions unfamiliar to him, he never hesitated to reveal a lack of knowledge, but would eagerly dive in with interest and curiosity to learn more, rather than immediately reject a new approach.

In a matter of months, Marcus's leadership style inspired members of his team who felt beaten down and rejected by other units to become more confident, challenge one another, and productively collaborate with internal clients. He was quick to affirm great ideas and reinforce good instincts, but he also never missed an opportunity to give critical feedback, especially when witnessing behavior counter to the culture he was building. If one of his staff members interrupted another colleague or was overly negative during brainstorming sessions, he would call out the specific behavior disruptive to the entire group. "Not okay" was his phrase and was picked up by others on the team. "Not okay" became shorthand for checking unproductive actions and promoting norms of what we think of as critical collegiality, the practice of encouraging colleagues to test assumptions and view problems through a different lens within the bounds of a trusting partnership.

In just over a year Marcus and his group successfully integrated and innovated numerous business analytics systems that supported a team that grew in the same period from a few dozen employees to 450. He is credited for building a business intelligence infrastructure that improved the timeliness and quality of decision-making and supported the company's successful public offering. When Marcus moved to his next professional opportunity, he left behind a system that could adapt and grow. But more importantly, he motivated a group of people who continued to lead as innovators.

It's hard to overestimate the importance of leaders modeling the other four innovation principles (Courage, Risk-Taking, Trust, and Collaboration) and challenging the behaviors that compromise innovation—lack of prioritization, inertia, and arrogance. It is when leaders signal to their teams that innovation is not only possible but necessary to genuinely improve diversity within an organization that Fixed Practices and Fixed Attitudes can dissipate, making way for real change to take root. Figure 3.3 illustrates the interplay between innovation principles and diversity.

The Virtuous Cycle of Innovation and Diversity

Innovation has long been a central driver to business growth; without it, entire industries become extinct.

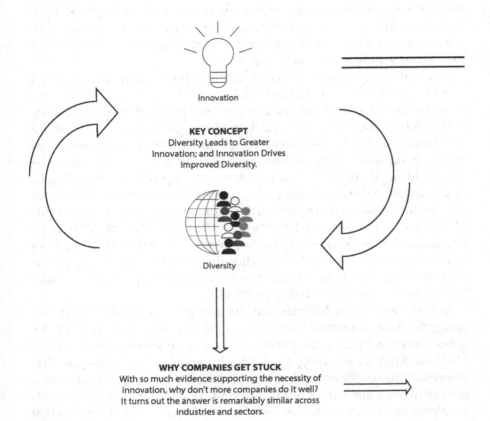

Innovation

KEY CONCEPT
Diversity Leads to Greater
Innovation; and Innovation Drives
Improved Diversity.

Diversity

WHY COMPANIES GET STUCK
With so much evidence supporting the necessity of
innovation, why don't more companies do it well?
It turns out the answer is remarkably similar across
industries and sectors.

Figure 3.3: The Virtuous Cycle of Innovation and Diversity, Illustrated

Five Cultural Characteristics
Necessary for Innovation to Thrive

To be **courageous** is to show
strength in the face of pain
or take an action that is
frightening or unpleasant.

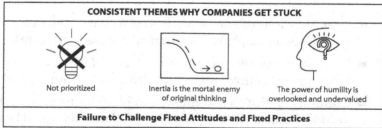

Leaders play an
outsized role in
creating conditions
where innovation
thrives.

Innovation represents
something **new** and
requires a departure
from what is safe
and known.

Collaboration happens
when people contribute
their unique experiences,
knowledge, and skills with
focus to achieve a common
goal.

Trust is predicated on mutual
respect, shared values, and
a belief that colleagues and
leaders will behave in ways
that benefit the goals of
an organization.

CONSISTENT THEMES WHY COMPANIES GET STUCK

Not prioritized

Inertia is the mortal enemy
of original thinking

The power of humility is
overlooked and undervalued

Failure to Challenge Fixed Attitudes and Fixed Practices

Figure 3.3: *Continued*

Conclusion

In this chapter, we examined the critical role of innovation in overall business success and summarized research documenting the positive influence of diverse teams on innovation outcomes. We also posited the idea that while diverse teams improve innovation, the reverse can also be true: innovation can be a tool to improve diversity. However, in order for a company to innovate—for diversity or any other business imperative—the culture must support Courage, Risk-Taking, Trust, Collaboration, and Leadership. Further, leaders must be vigilant against behaviors that dissuade innovation: lack of priority, inertia, and arrogance. When conditions and motivation exist for innovating for diversity, Fixed Attitudes and Fixed Practices fade away.

The innovation principles discussed in Chapter 3 will serve as a framework for case studies illustrating creative approaches to attracting, retaining, developing, and advancing diverse teams reviewed in the next chapters.

Summary

- Innovation is essential for business success; without it, companies and organizations become less competitive and put their survival at risk.

- Innovation is threatened when it is not a clear priority, organizational inertia is accepted, or when arrogant attitudes permeate the culture.

- Innovation benefits from diverse teams. Likewise, we posit that innovative practices can in turn drive and create diverse teams.

- Cultural conditions that must be in place for innovation to thrive include: Courage, Risk-Taking, Trust, Collaboration, and Leadership.

- When innovation is intentionally used as a tool for improving diversity, inclusion, and equity, Fixed Attitudes and Fixed Practices dissipate, paving the way to lasting progress.

Innovating the Apprenticeship Model to Advance Diversity in Tech

The apprenticeship model conventionally used in manufacturing and construction has, in recent years, been radically re-tooled to meet the growing demand for tech skills, and as a strategy for developing diverse talent. To re-imagine a centuries-old approach to tech requires a deep understanding of the populations to be trained, fearless leadership in developing and advocating a radically new model, and organizational commitment.

This chapter dives into the journey led by a small, innovative team at Citi to dramatically increase the number of veterans recruited into tech roles at their Irving, Texas, site. They developed an apprenticeship model after examining the limitations of earlier efforts, ultimately scaling it to a national recruitment and talent development program. We cover how they overcame challenges through external and internal partnerships, as well as through transparent communications and agile problem-solving.

Complementing the Citi case study, we share an example of how Accenture is using the apprenticeship model at scale nationally.

Key Concept: Passionate, driven leaders can tackle the challenge of hiring and developing diverse candidates like any other business problem—with the right support they can innovate, pilot, and scale effective solutions.

The Problem: Recruiting and retaining more tech talent from diverse and military backgrounds.

Citigroup, often referred to as Citi, is a preeminent banking partner serving institutions with cross-border needs, provides wealth management services globally, and offers personal banking services in its home market of the United States. Citi does business in more than 160 countries and jurisdictions, providing corporations, governments, investors, institutions, and individuals with a broad range of financial products and services.

Citi's success in attracting and developing U.S. military veteran talent, as well as individuals from diverse and non-traditional educational backgrounds, extends across its U.S. footprint. But the center of some of its most intensive hiring of diverse professionals is located just 30 minutes outside of Dallas, TX at its Irving site, where the company maintains significant data and IT operations. The facility has grown dramatically in recent years from a small service operation to over 10,000 team members in 2022. Technology staff alone numbered about 3,400 in 2022 and is expected to jump by a third to 4,500 by 2025. With this growth, Citi's demand for developers, cloud storage, and data security technologists, program managers, analysts, and administrators far outstrips the supply of local talent across North Texas. In the community, Citi is perceived as an excellent employer offering competitive salaries and benefits, flexible work arrangements, and is frequently featured on "best places to work" lists locally.

"It's time to get creative!"

In spite of its strong reputation and aggressive outreach efforts, the Citi HR and technology teams struggled to hire the number and quality of team members—veteran or civilian—to meet their goals. They knew they needed to get creative with expanding their sources of talent.

Citi has good working relationships with a number of nonprofit veteran-serving organizations, including Veterans on Wall Street, The Bob Woodruff Foundation, Green Extreme Homes, and the Wounded Warrior Project. One source of talent that Citi has tapped into for many years is the nonprofit tech training organization NPower. Citi has hosted interns and hired graduates from NPower's young adult and veteran program since 2012 for its New York City–based technical operations teams. In 2013, the company partnered with other employers and foundations to provide seed funding for NPower to open a class site in Dallas specifically to serve veterans. This decision was championed by senior leaders like Jon Beyman, who at the time led operations and technology for one of Citi's largest divisions, and who understood well the challenges of finding excellent employees for tech openings at all levels and across all locations.

The students accepted into NPower's inaugural year of training in Dallas—all of them veterans—began the program without significant prior technology experience but with ambition, determination, and a desire to improve their economic futures. The majority had enlisted in military service right out of high school or while in college in order to learn new skills, stretch their physical and mental boundaries, to help pay for further education, and out of a deep sense of duty and patriotism. Nearly all wanted a better future for themselves and their families.

During the 16 weeks of NPower's classroom training, students learned the basic fundamentals of technology infrastructure—including PC repair, networking, software installations, and security protocols—that prepared them to pass the rigorous A+ industry-recognized certification exam created and administered by the global IT professional organization, CompTIA. Even without a college degree, those with a CompTIA certification could land any number of well-paid, career-track tech jobs at entry-level and up. Beyond the technical instruction, NPower also tapped volunteer leaders from local companies, most of whom were themselves veterans, to provide the class invaluable advice on how to apply their profound military experiences to civilian tech careers. Problem-solving, critical and agile thinking under pressure, discipline, and collaboration are among the many skills service members learn and practice every day, but translating those skills within the corporate environment is not always straightforward. Empathetic coaching from someone who has shared experiences helps.

When NPower's first two cohorts of students in Dallas completed the technical training portion of the program, Citi agreed to host seven-week paid internships at its Irving site.

Every six months, with the cohort of new training program graduates, Citi would continue to tap a half dozen interns, about 75 percent of whom would be offered and accept full-time employment at the conclusion of their internships. Not bad, considering at the time employment conversion rates among college interns was about 50 percent across the industry. And, retention among those hired through the program was especially strong. Long-time Citi tech leaders spoke highly of the employees sourced through NPower and, overall, the program was working just fine as a pipeline for veteran talent.

The story could easily end here. The partnership led to about a dozen or so strong veteran hires every year—70 percent of whom were also underrepresented minorities. These hires thrived and grew as contributors and leaders. Most leaders would be satisfied and call it a day. Stuck in Fixed Practices and Fixed Attitudes, they would have believed they had done their best and let the program continue to carry on as a boutique project that made everyone feel good and delivered a few strong candidates a couple times a year.

Byron Cooper was one of Citi's very first interns in 2014 and one example of those strong candidates. Byron served four years in the U.S. Air Force as a medical service technician and began a career in transportation logistics and customer service as a civilian. He chose to enroll in NPower's IT training program after he was laid off in a merger between his employer and a competitor. He started as an intern analyst in Citi's Global Command Center and eight years later is a vice president for a unit responsible for the technology needed to process a half billion financial transactions globally every year. He recalls the "hiccups" in the early days of the program when there was little formal on-boarding, orientation, or structure. Because he had several years of corporate experience prior to enrolling in NPower, he had a jump on how to network within Citi and build the relationships required to become a full-time staff member and advance in the ranks. This skill proved an advantage over veterans with little to no civilian work experience.

"We can do better"

You'll remember the name Jon Beyman from earlier in the chapter. He was the NYC-based Citi tech executive who greenlighted the initial investment to launch NPower's training program in Dallas. He retired shortly after the decision but not before he saw the early success. Stepping into Jon's role was Rich Greenbaum, who was recruited to Citi by Jon five years earlier and who had a long and distinguished career leading technology divisions for Barclays and Lehman Brothers. As the global head of technology infrastructure at Citi, finding and retaining excellent talent—particularly those from diverse backgrounds—was a critical success factor in his role. He knew of NPower from Beyman's experiences and wanted to better understand what was working—and what wasn't—in the partnership. One of his biggest questions was, "why aren't we getting more talent from NPower in Irving?" especially since the candidates hired were doing well.

Rich is a talented technologist who also understands people and thinks deeply about the qualities of great leaders and what motivates them. His colleagues describe Rich as strategic, masterful at getting a lot done in highly complex bureaucracies, and an excellent developer of talent. In considering who might successfully assess how to best expand the hiring partnership with NPower to meet Citi's growing demand for technologists in Irving, he tapped Dan Maslowski, head of distributed engineered systems and storage who oversees a large operation at the Irving facility.

To say that Dan is an interesting guy is a little like saying Formula One racing is a nice activity. Not that you would know that working with him since he's pretty humble. It takes a little digging to understand why Rich would think he would be the right person to give the NPower program wings and to find any number of innovative ways to hire and support veterans for technical roles. Growing up a white kid in American Samoa, he'll tell you, when pressed, that he always felt like an outsider. His parents moved to Colorado when he was a teenager, and after graduating high school, without the resources to attend college, he took a job at McDonald's. For a time, Dan thought fast food could be his career. He traveled up the management ranks and attended McDonald's highly respected Hamburger University, where he came to appreciate Ray Kroc's belief that "none of us are as good as all of us." But as a young manager in his early twenties, he thought about the road ahead and decided regional management of a fast-food chain in his forties was not the path for him. Instead, he enrolled at the New Mexico Military Institute, where he earned his associate degree and began service in the U.S. Army.

Dan talks a lot about people who gave him a chance. Like many of his peers, he served in the military. His first civilian job was as a security guard, a role he excelled at, but the pay wasn't great. He shares the story of walking the night shift in shoes lined with cardboard to protect his feet from holey soles. The manager who hired him for the security job at a large hotel, Dean Hazlewood, observed his strong interpersonal skills and suspected he had the potential to do a whole lot more but needed to hone his professional skills. Dean offered him an opportunity to work in the events department where he learned consultative skills and got a chance to use his audio-visual and tech skills, a side passion of his for years. In fact, it was from this experience that he decided to focus the next stage of his career solely on technology.

Dan earned his undergraduate degree in computer systems through a part-time program and took his first real tech job at a computer storage company known at the time as Adaptec. Again, Dan noted that the hiring manager, a former petty officer in the U.S. Navy, acknowledged that he didn't have the traditional background Adaptec usually looked for but took a risk and gave Dan a six-month contract to "kill it." He worked hard, succeeded, and stayed with the company for nearly five years, becoming a software engineer. From there, Dan progressed through the ranks at Sun Microsystems, AMD, and Oracle before joining Citi. He views his career as a series of transitional moments when others saw more in him than he saw in himself. In turn, he is filled with enthusiasm for providing opportunities to others whose paths haven't always been clearly marked.

Rich Greenbaum knew Dan was the ideal person to figure out how to build the partnership with NPower Texas, and he was right.

What Needed to Change?

Dan pursued the challenge of sourcing more tech talent from NPower like he would any technical or business problem. He enlisted respected colleagues, Bobbie Long, senior vice president and global engineering lead, and Raymond Pitts, senior vice president of global data center strategy, to collaborate on solutions. Both Bobbie and Ray—each with their own affiliations with the veteran community—were key partners in the process. Together they spoke to NPower graduates working at Citi and their supervisors, spent time with the NPower team locally in Dallas, and formulated a clear set of strengths and weaknesses of the program.

What was working:

- Veterans from the 16-week NPower training program entered the Citi internships with solid fundamental technical skills and respected industry certifications, which helped them get up to speed in their assignments.

- The interns had a good work ethic and were eager to pick up new skills—they were already in "learning mode."

- There was a strong camaraderie among the cohort of interns coming in together. The encouragement they provided for one another built confidence.

- NPower staff continued to be engaged, offering continued professional development and coaching support when needed.

- The team at Citi enjoyed working with the interns individually and as a group. The supervisors involved offered up excellent general feedback on the caliber of the interns.

- Funding to host the interns came from a separate operating funds managed at the divisional level, so local leaders did not have to tap their own personnel budgets. This removed a potentially significant barrier.

What wasn't:

- The seven-week internship was not long enough for the intern to complete projects, or for supervisors to fully assess potential for full-time employment. As Bobbie noted, "it was barely enough time to find your way around the place."

- Many of the interns still needed additional specialty training that complemented the technical skills they learned at NPower to be fully productive in their roles. They also needed support in navigating Citi's complex corporate structure and unique vocabulary.

- Because the number of interns brought on was small, the effort appeared to some Citi leaders to be more of a one-off without sustained commitment or a longer-term plan. It was unclear to managers just how much personal time to invest.

- Resumes of the NPower intern candidates were managed through the same process as college hiring, and sometimes overlooked as qualified applicants.

- This was not a formalized, traditional college internship program that worked on a typical summer schedule. NPower interns were ready for conversion to full-time employment immediately following completion of their assignment. The timing and steps to hire were not consistently transparent to the supervisor or to the intern.

- The majority of the NPower interns, while all veterans, came from very different educational, racial, and socio-economic backgrounds than the full-time Citi staff. Many participants believed they didn't fit in and struggled to understand the culture.

- Most of the interns were keen to improve their economic situations, which is why they sought out the NPower program in the first place. For the duration of the internships, participants were paid about 20 percent more than the federal minimum wage. Financial anxiety, uncertainty over whether full-time placement would be forthcoming, and cultural dislocation combined to prevent interns from believing they were performing at their very best. Bobbie relayed the story of a veteran who moved to Texas specifically to join the NPower training program, which he successfully completed and was chosen to intern at Citi. He showed up every day ready to work and very quickly stood out as capable and motivated although, from time to time, tired. Only when he was offered a full-time position, did he reveal he had been living in his car. That scenario lingered with Bobbie and the team and deepened their resolve to constructively support the interns.

The Solution

Bobbie, Ray, and Dan concluded that by improving the experience for both the interns and the supervising managers, they could meaningfully attract a greater number of participants, advance their performance, and increase the percentage who converted to full-time employment. Taking such steps, they figured, could more than double the size of the NPower talent pipeline in one year, at a time when hiring needs were greater than ever. Working together, they brought more structure to the program; engaged vendor partners; tapped the energy, passion, and expertise of Citi Salutes, their veteran-focused employee resource group (ERG); and found ways to better integrate the interns into the culture of the Irving operation.

Adding Structure

Dan reflected on his own early days of onboarding at Citi. The manager who recruited him, Greg Lavender, advised, "Dan, good engineering skills alone won't cut it at Citi. You need to build relationships, work well horizontally, understand how the place works, and positively contribute to the culture." Greg gave Dan nearly twelve months without mission-critical assignments, so he had the time to shadow other leaders, learn the strengths of others around him, understand how decisions were made, and internalize the cadence of various projects. By the time he interviewed at Citi, Dan was already a seasoned tech executive. He reasoned, "If *I* needed a year to really figure out the job, how can we expect these veterans—new to tech and most new to corporate culture—to really succeed in just seven weeks?" He talks about how Greg "removed the fear" during his first year at Citi and gave him permission to fail, within reason, and helped him find his stride. He believed he had quicker and higher-impact successes because of Greg's decision to invest both managerial attention and time during his early days at the company. In turn, he believed the interns deserved a similar investment. Bobbie had come to a similar conclusion and she and Ray both supported the idea.

Dan presented a new plan to Rich Greenbaum, the global head of technology infrastructure, who tasked him with finding improvements several months earlier. Dan suggested that the seven-week internship period get extended to a full six-month contracting assignment and asked for an increase in budget to accommodate the additional time. With Bobbie and Ray's counsel, he recommended other changes so that the apprentices would:

- Improve their chances for getting selected into the program
- Benefit from additional technical training relevant to their roles
- Receive a higher salary, bringing them closer to competitive compensation for full-time entry level talent in the market
- Know where they stand on the likelihood of an offer to join Citi full time at the conclusion of their assignment
- Receive regular performance check-ins with their supervisor
- Get formal on-boarding time to become familiar with Citi systems and procedures
- Be assigned an NPower alum who was also a current Citi staff member as a peer mentor
- Actively engage in Citi Salutes ERG events and networking, which also presented opportunities for the apprentices to better understand the Citi culture

In reviewing Dan's recommendations, Rich considered several qualitative and quantitative factors, including the frequent, enthusiastic comments he received from Irving-based colleagues eager to support the veteran community and who volunteered positive feedback. Importantly, even with the increased cost of the higher salary, the program was still an efficient way to bring qualified, diverse talent, and the participants could productively contribute during their six-month tenure. Rich also suspected that by enhancing the program, Citi could further develop its reputation and become the Number One destination for tech talent in North Texas. And finally, like Dan, he believed the program was "the right thing to do" and contributed toward a more inclusive and welcoming culture. Even in the early days, employees were galvanized around a program that benefitted the business and supported talented but overlooked veterans who left military service with a track record of bravery and sacrifice but few civilian opportunities.

Dan often referred to Rich's support and direction as the air cover he needed to drive a new program, especially when others would question the time and energy required or challenge the qualifications of candidates. From Rich's perspective, he was supporting an effort that expanded the pipeline of good candidates and also helped advance Citi's global goals to build a thriving, diverse workforce. The way Rich put it, "I wanted to add value to Citi while creating opportunities for others too often overlooked."

Engaging Partners

The first step was to formalize a six-month program that included new components designed to support the long-term success of the participants. As luck would have it, at about the same time Rich agreed to extend the internship window, NPower had received approval from the U.S. Department of Labor to become a Registered Apprenticeship Program (RAP) in the areas of IT generalist and cybersecurity. Traditionally in the U.S., apprenticeships have been associated with skilled trades, such as carpenters, electricians, plumbers, and pipefitters. However, with the shortage of tech workers, the Department of Labor has increasingly added funds to create programs in non-traditional categories, with good success. To become a registered IT Generalist apprentice, students were required to complete formal class training and gain relevant on-the-job experience. The class time with NPower that prepared the students for their CompTIA A+ certification test satisfied the formal training requirement.

World Wide Technology (WWT), the large St. Louis–based technology services provider, was among NPower's very first apprenticeship employer partners. After completing the classroom instruction, NPower students in St. Louis were then hired by WWT to complete the on-the-job training requirements, getting paid along the way. They learned additional skills such as project management and process improvement methodology. Upon completion of the training period, over 90 percent of apprentices became full-time WWT associates.

WWT also happened to be an important technology systems partner of Citi. Chris Young was a lead WWT client manager working with the Citi account in Irving. He worked closely with Bobbie as a client and shared Citi's interest in extending the NPower interns to six-month assignments. Like Dan, Chris enlisted in the U.S. Army National Guard, and now serves as an officer of the Army Reserves. He is passionate about helping veterans find meaningful employment in the tech sector. He and Bobbie brainstormed a clever idea: Why not give the NPower interns in Texas the opportunity to become nationally recognized IT apprentices during that six-month period? In fact, WWT could become the employer of record and handle all the Department of Labor apprenticeship paperwork, as well as provide additional training, but place all of the apprentices on site at Citi's Irving location. The NPower participants would have the benefit of Citi onboarding and mentorship, as well as access to WWT's innovative Advanced Technology Center, where they could develop a range of new tech skills and learn from subject matter experts. As the apprenticeship employer partner, WWT also supported participants with earning additional in-demand certifications. At the end of the six-month on-the-job apprenticeship period, Citi could exercise an option to convert any of the apprentices to full-time staff with no fee. If Citi chose to take a pass, WWT would find another role for any of the apprentices performing well.

Improving the Apprenticeship Selection Process

Bobbie Long zeroed in on the application process to identify veteran interns—the same process used for hiring recent college graduates—and understood that good candidates were missing out. She set in motion a new way to identify candidates by going directly to the source. She and Chris Young, the WWT client manager, met with the NPower team to explain the top characteristics they would be evaluating. NPower supplied resumes that both Bobbie and Chris personally reviewed, then set up "speed dating"—like interviews with managers at Citi with the fifteen or so candidates they selected. Managers had an opportunity to rank their top candidates. And, similarly, candidates ranked the tech specialty areas they would most like to gain experience in. Bobbie and Chris, together with a key leader from the diversity team, Rebecca Moss, selected ten who were matched as closely as possible to mutual preferences and hired on by WWT for the six-month apprenticeship period.

First Pilot Outcomes

Dan and the Citi team piloted the arrangement with WWT with ten IT apprentices with great success. In fact, Dell Technologies, a partner to both Citi and WWT, learned of the partnership and offered to add in specialty training on cloud storage and data security during the apprenticeship period. At the conclusion of the program, nine of the ten apprentices were hired by Citi, and one became a full-time WWT associate employee.

In addition to fulfilling Dan's plan to extend the on-the-job period to six months, the partnership with WWT also checked off the first four of Dan's recommendations—offer additional skills training, improve compensation, clarify the job offer timing and process, and provide regular performance feedback. Still, there was more to be done with the support of Citi's internal team.

Even with WWT serving as the employer of record, the apprentices still needed to learn their way around Citi culture and expectations. Since conversion to full-time status was the goal, Citi and WWT agreed that the apprentices stationed on site would be immersed in Citi's way of doing business. To create a rewarding experience for the apprentices and support their professional success, Dan, Ray, and Bobbie, together with Chris, next focused on the remaining three recommendations: effective on-boarding and coaching, peer mentorship, and engagement with Citi's veteran ERG group.

The Importance of Consistency to a Good Start

When the NPower veteran internship program first started at Citi, it was a very decentralized process. The NPower team would send the HR department resumes of veteran candidates for the seven-week internships, and it was up to individual managers to review and select, if anyone, for those slots. Once brought on board, each manager decided how to on-board and train the intern.

The start of the formal apprenticeship program with WWT was an opportune time to change this decentralized approach. To create a more consistent on-boarding, introduction, and training experience, Bobbie Long volunteered to become a "central manager." In this role, she ensured all apprentices received a proper Citi orientation early in their tenure. She also met weekly with each apprentice, which allowed her to get to know each individual personally, and understand their strengths, weaknesses, and what was working (or not) about the apprenticeship. She had a central manager counterpart at WWT, Scott Ream, and together they closely followed the progress of each apprentice. Bobbie also played another important role during this time. She provided needed coaching to direct supervisors who had never managed veterans. Bobbie was an ideal leader for this job. Numerous members of her own family served in the military and, over the years, she took on the responsibility in her past jobs with other companies to help veterans and managers successfully translate "military speak to corporate speak."

The Role of Battle Buddies

The Citi team—and NPower graduates now working at Citi—will point to the idea of "Battle Buddies" as one of the keys to the success of the program. Each new apprentice was matched to an alum from NPower working full time at Citi who could serve as a peer coach and help unlock organizational complexity, much as Greg did for Dan years before during his own transition. Iris Fagan, an army veteran who specialized in environmental health during her military service, joined Citi full time as a business analyst. She reflected on the importance of her battle buddy, Cecile Comartie, during her apprenticeship. "Cecile helped me with the small things, like decoding the hundreds of acronyms Citi uses, and showed me how to navigate the Citi structure, especially when it came time for offers of full-time jobs. She's still my buddy." Another apprentice, Steven Guiliani, a veteran of the U.S. Marines, described his own on-boarding process as "somewhat chaotic and complicated" and committed to supporting others by volunteering as a battle buddy and strengthening the on-boarding process to ease the way for others. Empowering alumni from the program to "pay it forward" contributed to a culture of camaraderie and loyalty. When asked, "where do you see your career headed in the next 5–10 years?" all the former apprentices without exception described a future where they grow in responsibility with Citi.

The Citi Salutes Impact

Ray Pitts played an important role in closing and opening Citi data centers across the U.S. and Latin America. He learned a wide range of telecommunication, satellite technology, and software engineering skills in the 13th Signal Battalion, having served in the U.S. Army for 16 years. He was a key leader in organizing the activities of Citi Salutes, Citi's global employee resource group supporting military service members, veterans, and their families. Ray first got to know NPower when he was invited to speak at one of its classes in Dallas. He took a personal interest in the NPower veteran apprentices and spent time getting to know individuals and their stories. Not only did he invite the apprentices to join Citi Salutes networking events and volunteering projects, but he encouraged members of the Citi Salutes group to share their own backgrounds and their journeys to civilian careers. Through these personal exchanges, the apprentices tapped into the spirit of solidarity that many felt was so rewarding during their time on active duty. Being welcomed into the Citi Salutes circle helped to validate for the apprentices that they belonged, even though many did not have college degrees and almost none had previous corporate tech experience. They were all held to high standards, but over six months, the participants came to understand and appreciate that the non-tech skills they learned in the military—discipline, resourcefulness, leadership, ability to work under pressure, problem-solving—were just as valuable to Citi's Irving operation as their newly acquired tech certifications.

MALCOM SMITH: AN EARLY APPRENTICE HIRE

Malcom Smith is vice president of Citi's business continuity group, which is responsible for technology recovery. When the pandemic struck, his business unit was responsible for quickly getting people online into secure systems so they could be productive from home. He often responds to emergencies, and it wouldn't be unfair to call the job hugely stressful. Yet you would be hard-pressed to find a leader more spirited, energetic, and upbeat about both life and work than Malcolm. He is in exactly the right role at the right time.

Malcolm describes his journey as being one of "firsts." He was the first in his family to attend and graduate college, enlist in the military, and pursue a career in tech. Malcom joined Citi in 2018 through the apprenticeship program with WWT after serving in the U.S. Navy as a weapons specialist for nearly eleven years, with several deployments to the Middle East. He remembers his frustration after sending hundreds of resumes to employers once he left the military and getting few responses, "Veterans are assets. We can do so many jobs; we have leadership, work hard, and can learn and adapt quickly. We just need a foot in the door!"

His supervisor at Citi once remarked how well Malcolm performed under pressure. To Malcom, nothing on the job couldn't be tackled as a problem to solve. Thinking about both his childhood growing up in a tough Milwaukee neighborhood and his deployments to Afghanistan, "I don't panic when emergency situations arise at work. We're not getting shot at, not getting the lights turned out, and we're not going hungry. There's not much we can't tackle here."

Malcom was recently promoted and sees a long and successful career for himself at the company. "I love my job. I want Citi to do great—if Citi does great, I do great and we can all do great things together." He sparkles when talking about serving as a role model, supporting others who come from veteran and underrepresented backgrounds, and paying it forward. And adding value to Citi.

It's hard to imagine any employer who wouldn't be thrilled to have Malcom's intelligence and leadership in their ranks. Citi had the vision and understanding to look carefully between the lines of a military-centric resume that other employers dismissed, or just didn't have the patience to de-code.

Glitches Along the Way

As with any new endeavor, not every aspect of the program worked well with every cohort of apprentices. As technologists, it was natural for the team to adopt agile methodology in reviewing problems and quickly correcting course. Dan, Bobbie, and Ray met regularly as a team, together with Chris Young and other colleagues at WWT and Citi, to review observations and performance. They were all committed to the common goal of maximizing the opportunities for the success of each individual apprentice and creating a long-term and sustainable pipeline of diverse veteran talent for Citi's Irving operation. With agreement on the vision and trust in one another, no one member of the team held back in openly discussing challenges or constructively engaging in solutions.

Unsuccessful Recruitment Choices

Over the next two years, Dan and his Citi colleagues continued to make refinements in the selection criteria of apprentices, the partnership with both WWT and NPower, and the structure of the program. Not every apprentice completed the program, and not everyone who did was hired full time. Generally those who dropped out either decided tech wasn't for them after all, or realized the Citi culture was not a good fit for them. In either case, the NPower career placement team was successful in finding those candidates other employment; having four to six months of Citi tech experience on a resume didn't hurt. After a few cohorts, the WWT, Citi, and NPower recruiting teams together formulated clear profiles of prospective apprentices that highlighted the aptitudes and characteristics most likely to thrive in the program.

Supervisor Limitations

WWT played an important role in providing oversight for the performance of the apprentices, but onsite Citi supervisors still needed to be actively involved. While Citi supervisors were given orientation and guidance, not all were able to successfully nurture the apprentices in the program. Those supervisors were removed from consideration as future beneficiaries of apprentice talent. Over time, Citi supervisors were self-selected by building superior track records of developing apprentices into successful full-time hires and welcomed the opportunity to augment their teams with talented candidates. Interestingly, as the program became more and more successful, some supervisors who initially weren't as successful with the program asked to re-engage after taking the time to learn how to effectively manage the veteran apprentices.

Skill Mismatches

From time to time, the process of matching apprentices with supervisors and job roles just didn't work. The great value of Bobbie serving as a "central manager" was that she could readily spot when an individual wasn't quite right for the job assigned. She recalls a time when an individual was placed in a department requiring significant coding abilities. He was struggling, but she noticed that he had excellent project management skills. He was quickly transferred to a new area where his strengths would shine. He was brought on in a full-time role and, several years later, is described as a "huge contributor." Bobbie reflected somberly on how, in many circumstances, the story of this apprentice would have ended differently. He could have been dismissed as "not a fit" and left to question his abilities and his potential for a future tech role anywhere else. She added, "these apprentices are not just numbers, they are real people with aspirations, talent, and potential."

And, in the end, this is perhaps one of the most important lessons of the Citi-WWT partnership relevant to almost any endeavor to improve diversity, equity, and inclusion:

- Focus on the individual.
- Get to know and understand unique skills and interests.
- Create a culture and environment that supports success.
- Find passionate and determined leaders energized by unlocking potential in others.

APPRENTICESHIPS AT SCALE

As Fortune 100 companies increase long-term investment in enterprise technologies such as digitization, cloud migration, data management, and cybersecurity, demand for Accenture's services, and those of other external consultants, continues to climb, especially as client companies struggle to meet their own internal technology staffing requirements in the face of a rapidly changing digital economy.

The challenge for Accenture is building its skilled workforce fast enough to keep pace with client demand and accelerate its capacity for innovation and growth. By the second quarter of 2022, Accenture employed about 700,000 people globally, up nearly 200,000 in just two years. Accenture takes its obligation seriously "to deliver on the promise of technology and human ingenuity" and commits roughly $1 billion yearly to upskill its fast-growing workforce. Julie Sweet, Accenture's chair and chief executive officer, frequently speaks on the value of learning agility across any organization. She is known for regularly introducing training topics at leadership meetings and personally commits to her own learning objectives every quarter. Accenture considers the ongoing, rapid training of employees – from entry-level staff to the C-suite – a core competency.

How and Why the Program Began

Accenture's apprenticeship program was first launched in Chicago in 2016 and as of September 2022, has since grown to 2,000 apprentices across North America, meeting the company's goal to fill 20 percent of its entry-level roles from its apprenticeship program for its fiscal year 2022. The majority of apprentices are racially diverse, nearly half are female, and the vast majority joined the company without 4-year college degrees. They successfully work in client-facing roles as well as internal functions.

Pallavi Verma, senior managing director and executive sponsor of the apprenticeship program at Accenture, recalled the early days of the program when she ran the Midwest region: "Mayor Rahm Emanuel challenged the Chicago business community to hire diverse talent and to recruit from community colleges. Through the apprenticeship program, we tackled the challenge head-on."

The apprenticeship program in Chicago started small, with five apprentices. Nearly seven years later, three of the five original apprentices continue to grow with Accenture, and one left for another opportunity after thriving at the company for six years.

Danica Lohja was among the first cohort of apprentices selected in 2016. She joined after receiving an associate's degree from the City Colleges of Chicago. She came to the U.S. on her own from Serbia in 2011 in search of better opportunities and quickly found work as a waitress, later as a tutor and an administrator for an appliance and electronics superstore working on vendor contracts. Danica learned of Accenture's apprenticeship program from an email forwarded to her by the City Colleges of Chicago career center and was encouraged to apply. After an office visit and interview with Accenture, she was elated to learn she was accepted into the program. In her first role, she was responsible for managing a tool for tracking supplier relationships, which built on previous skills she learned while working at the electronics store.

Meeting the other members of the first cohort of apprentices, she recalls, "None of us knew quite what to expect, but we did know we wanted to make the most of the experience." Thinking back on the first few months of the program, Danica shared, "We were given real responsibilities from day one. About half of our time was spent in relevant technical training, and there was balance in performing our responsibilities, job shadowing, and internal networking to learn the culture and business."

Danica was thrilled to have a supervisor who invited her to various meetings, after which he would always ask her, "what two questions do you have from this meeting?" He encouraged her to listen and observe carefully and spent time discussing what she had learned.

Six years later, Danica is an associate manager at Accenture in a role managing the "360 relationships" of important vendor and client partners. She has been promoted three times and sees an exciting future ahead for herself at the company. With a personal understanding of the power of the apprenticeship initiative, Danica regularly volunteers to serve as a mentor or "apprenticeship buddy" to those coming after her. In getting to know each new cohort, she observed, "they are all eager to learn, to develop, and to grow with Accenture."

The apprenticeship program contributed to the strong culture of mentorship at Accenture. Colleagues in the Chicago office were quick to mentor the apprentices and were invested in their success.

Overall, the overwhelming majority of apprentices who complete the program stay on with Accenture, with continuing opportunities for long-term career growth. Through the "learn and earn" model of the apprenticeship program, Accenture apprentices receive market-based wages and benefits as they prepare for in-demand roles, including cybersecurity, digital, data analytics, and cloud migration.

Building off the success of its internal Apprenticeship Program, Accenture joined forces with Aon in 2017 to form the Chicago Apprentice Network with the goal of hiring 1,000 apprentices city-wide by 2020. By 2022, the Chicago Apprentice Network surpassed its initial goal, with more than 1,300 apprentices working at more than 75 employer partner companies throughout Chicagoland.

Success Factors for Accenture's Apprenticeship Model

Pallavi and Wendy Myers Cambor, managing director and talent strategy lead at Accenture, shared several reflections on what makes the company's apprenticeship model successful:

- **Leadership needs to be unambiguous** – Bold goals established by the CEO send a clear signal that diversity and alternative strategies for identifying and developing talent, such as apprenticeships, are a priority.

- **It's not just the job of Human Resources** – Commitment to hiring diverse talent from previously untapped sources through apprenticeships or other means cannot be the sole responsibility of the chief human resources officer (CHRO). Rather, it must be a close collaboration between HR and business leaders who commit to supporting the success of the new hires.

- **Old ways of recruiting must be disrupted** – Evaluating candidates through the traditional one-page resume focuses on narrow educational credentials while focusing on "skills first" and experience means great talent is not overlooked. Accenture re-tooled its recruiting practices and developed a specialized recruitment team dedicated to identifying potential over pedigree, successfully sourcing previously overlooked talent for the apprenticeship program.

- **Collaboration expands community reach and talent pipelines** – Accenture has forged a network of over 200 partners nationally, including community colleges and nonprofit and for-profit training providers that have become robust sources of apprenticeship talent. The specialized recruitment team builds partnerships with community-based organizations similar to those historically established with four-year institutions to recruit college graduates.

- **Apprentices are the best ambassadors** – Accenture encourages employees who joined through the apprenticeship program to share their experiences and personal stories to encourage others to apply and to inspire business leads within the company to hire apprentices. Wendy notes that whenever Accenture hosts a company event featuring a speaker from one of the apprenticeship cohorts, hiring managers are quick to call her for candidates.

Like we saw with the Citi case study, the personal journeys of those who are too often overlooked as viable candidates for career-track professional roles can sharply re-frame what it takes to succeed in an organization – prioritizing skills, hard work, ambition, and agility. Those stories also motivate passionate leaders to provide mentorship and coaching in ways that contribute to a culture supportive of the success of all employees who demonstrate core values and consistent performance.

Pallavi and Wendy both credit the apprenticeship program for helping Accenture continue to thrive during the pandemic while other firms struggled to find qualified talent. As Wendy says, "It is a rare example of innovating at the very intersection of business strategy, talent strategy, and social impact strategy."

Two-Year Outcomes at Citi

The original goal was to double the number of full-time staff members recruited from NPower for the Irving operation, which would have been about 24 over a two-year period. The team far exceeded that goal. Within two years, the Citi-WWT-NPower partnership brought on board 45 apprentices with a 90 percent conversion rate to full-time employment. All of those participating in the program were veterans or the spouses of veterans, and over 70 percent were Black or Hispanic. Individuals were hired into in-demand roles as Program Manager, Program Coordinator, Service Desk Analyst, Middleware Engineer, Operations Support Specialist, Junior Storage Engineer, Systems Engineering Analyst, Business Analyst, and Partner Support Manager.

Over time, the team continued to fine-tune the program, which led to even better candidate selection, full-time conversion, and excellent retention levels. And helped supervisors to become more effective managers of veteran talent. The engagement of members of the Citi Salutes™ ERG and Battle Buddies got formalized into what is today known as the "Pathfinders Program" designed to help 100 percent of the apprentices succeed as long-term employees at Citi. With this more structured program, other leaders across the Irving operation serve as advocates and mentors.

Ray observed that in-depth one-on-one coaching with the less outgoing or less confident apprentices helps them build better relationships with their direct supervisors, request feedback, and be fearless in asking questions about assignments. In turn, the Pathfinders Program provides peer coaches to help supervisors interpret and respond to behaviors more likely demonstrated by veterans than civilians. For example, some supervisors perceive veteran talent as more formal or, as Ray put it, "straight up and down" and reluctant to spontaneously engage with those who are more senior, which can make it a challenge for supervisors to get to know them. Over time, thanks to coaching, most supervisors have come to understand the nature of veterans' respect for hierarchy and now know how to encourage more relaxed interactions.

From Pilot to Operationalizing: Expanding Irving Success across Citi

As the Irving pilot expanded in 2019–2020 and became more successful, word traveled quickly across Citi, and other divisions expressed interest in participating. At about the same time, Rich Greenbaum retired from Citi, and his colleague Jennifer Kleinert volunteered to be the new senior-level champion for cultivating and expanding the partnership with NPower. Jennifer is the Chief Operating Officer for Citi's largest operations and technology division and brought to the partnership her abundant energy, incisive thinking, and "get it done" attitude. As a broad systems-wide thinker, she looked at the program with a fresh perspective and saw an opportunity to "mainstream" the apprenticeship approach to a number of other units. Her first decision was to bring the program in-house. While Citi maintains a strong partnership with WWT for a range of other services and functions, Citi now brings aboard apprentices directly. Her second decision was to build in access for the apprentices to many of the same training and development opportunities reserved for college interns and recent graduates. She had a capable partner in Rebecca Moss, a key member of her Business Office team, who played a vital role in managing the program once centralized.

Interestingly for WWT, the deep experience they gained in forging an apprenticeship program with Citi has evolved into new service offerings to support other WWT clients and represents innovation borne out of a partnership originally designed to expand a diverse talent pool.

By 2022, *Citi had hired over 200 diverse veterans and young adults* from NPower's training program for technology apprenticeship positions across multiple parts of its business, including compliance and control functions as well as its cyber intelligence and cybersecurity fusion centers. Jennifer expects that annual hiring will accelerate as more Citi divisions find success with NPower graduates and as NPower scales to meet demand.

While the quantitative measures are impressive, Jennifer is quick to offer up qualitative benefits of the apprenticeship program to the overall culture at Citi:

- Apprentices are eager to learn and have an undeniable "fire in the belly" (an expression used by many at Citi in describing both young adults and veteran participants). Their appetite for achieving inspires peers and managers around them to stretch further.

- Apprentices are quick to see the range of future opportunities at Citi and remain with the company with high levels of retention well beyond two to three years.

- The program has produced a "trickle up" effect where the enthusiasm of the apprentices infuses the managers and mentors with renewed energy, commitment, and loyalty. Every leader interviewed highlighted this benefit with one commenting, "I wake up every day excited to see the progress of the apprentices." Bobbie and Dan separately noted that the demands of their "day jobs" didn't abate while they got the program off the ground but found purpose—and therefore the time—to make it a success.

▪ The culture has shifted to consider and embrace other partnerships for attracting diverse talent, including a new collaboration with the Information Technology Senior Management Forum (ITSMF), an organization committed to supporting Black technology professionals in leadership roles.

▪ Human resources has become a critical partner engaged in the national "operationalization" efforts and focused on measuring longer-term impacts.

For NPower's part, to meet increased demand from Citi and other partners it is scaling to new markets. It's also expanding a SkillBridge program with the Department of Defense to train service members who are six-months from completing their military service and transitioning to civilian careers.

How does the Citi Case Study stack up against the Innovation Principles Framework?

Let's review how the Citi's approach to inventing and operationalizing a program to recruit and develop veterans reflects the innovation concepts discussed in Chapter 3.

Aligning Apprenticeship Programs with Innovation Principles

PRINCIPLE NECESSARY FOR A CULTURE OF INNOVATION	PRINCIPLE IN ACTION
Courage	▪ Fearlessly questioned elements of the original veteran internship program that weren't working.
	▪ Challenged and changed the existing HR process for reviewing applicants from veterans who had technical certifications, but often not a college degree.
	▪ Asked for additional budget to extend seven-week internship to a six-month apprenticeship.

PRINCIPLE NECESSARY FOR A CULTURE OF INNOVATION	PRINCIPLE IN ACTION
Risk-Taking	▪ Expanded and invested in a program that sourced talent with non-traditional educational backgrounds, a departure from the norm of recruiting recent college grads.
	▪ Air cover offered by senior leadership created environment supporting inquiry and testing.
	▪ Potential reputational damage, especially with significant personal commitment and momentum invested.
Trust	▪ Colleagues in Irving formed a partnership based on trust, respect, and a personal connection to the goal of helping veterans succeed.
	▪ Mutual commitment to the goal facilitated honest, direct exchanges that led to continual improvement.
	▪ Senior leadership in New York invested local Irving leadership with the confidence—and expectation—to land on a smart solution.
Collaboration	▪ Expansive mindset led to a boundaryless solution that relied on trusted external partners that each played defined roles.
	▪ Structured formal networks within the company to help civilian leaders understand and leverage into the unique strengths and characteristics of veteran talent.
Leadership	▪ Tapped into the personal, lived experiences of key leaders to produce creative solutions.
	▪ Demonstrated by example the power of personal conviction, a desire to "pay it forward," and an unwavering belief in the potential of veteran talent.
	▪ Fostered an environment of creativity and safety while setting clear expectations.
	▪ Effectively challenged what for some was a Fixed Attitude around hiring and managing veteran talent.
	▪ Enhanced and scaled a successful local program for national impact.

Innovation Threats and Preventative Actions

INNOVATION THREATS	ACTIONS TO AVOID THREAT
Low Priority	Clear goal to double the number of hires through the program, combined with the team's sense of purpose and mission, prevented work on the program from becoming a "bottom of the list" afterthought. Steady progress kept team motivated, even though their work on the project was additional to their regular responsibilities.
Inertia	Senior leadership asked the right question—"Why can't we do better?"—that prevented a just-adequate program from becoming the acceptable standard.
Arrogance	When an apprentice didn't work out for a particular role, the team had the humility to question their decisions in the matching process rather than blame the potential of the apprentice. Rather, they reassigned the apprentice to a role better suited to their skills.
	While the team had their own connections to the veteran community, they didn't assume answers based on their own experiences, but actively sought the opinions and expertise of others when building the program.
	The team met weekly to relentlessly examine their progress, openly confront problems, and applied an agile methodology to continuous improvement.

Conclusion

The experience of the team at Citi in building a thriving, tailored apprenticeship program proved how a novel approach could meet twin objectives of improving the pipeline of skilled talent and substantially increasing a source for diverse veteran employees. Central to the success of the program was a clear leader who was personally passionate about both goals, together with a committed team dedicated to "paying it forward" within the veteran community while adding business value to Citi. The team benefitted from the support of senior executives who created the conditions for innovation and partnerships with external organizations willing to flex to help Citi meet its goals. What began as a small pilot for one operation grew over time to a viable, consistent source of diverse talent for multiple technology roles across Citi nationally. The success of the team was in no small part a result of a strong culture of innovation and their adeptness at intentionally avoiding the three traps that stall creative approaches: low prioritization, inertia, and arrogance. They also avoided getting stuck in Fixed Attitudes and Fixed Practices by fearlessly questioning what wasn't working in the early internship program and risking personal and professional capital to advance new solutions.

As a sidebar, we also highlighted the work of Accenture to recruit at least 20 percent of their entry-level roles in North America through a structured apprenticeship program designed to attract diverse candidates. The initiative has the potential to reshape how apprenticeships are used to skill diverse technology talent across industries.

Summary

- Driven business leaders personally invested in finding DEI solutions are pivotal to successfully galvanizing partners and colleagues to seed change.

- Great leaders don't settle for "good enough" when it comes to identifying and training diverse talent. They push their teams to question what's possible and to innovate scalable solutions.

- The centuries-old practice of apprenticeships, traditionally deployed for construction and manufacturing roles, can be retooled as an effective strategy for recruiting and developing diverse technology talent.

- DEI should be approached as any other business problem to solve; the use of agile methodology to constantly improve Citi's apprenticeship program for veterans serves as an example.

- An apprenticeship program can not only professionally advance the individuals recruited but can also have a "trickle-up" benefit on incumbent staff members engaged in the program leading to a deeper sense of purpose, loyalty, and a reinforcement of an inclusive culture.

- Successful DEI practices *(1) focus on the individual, (2) understand employees' unique skills and interests, (3) create a culture and environment that supports success, (4) leverages passionate and determined leaders who are energized by unlocking potential in others.*

Creating High-Impact Mentoring Programs

Mentoring programs can potentially yield benefits to employers and employees alike. For participants, or mentees—particularly those from diverse backgrounds—they lend the opportunity to learn new skills, develop their professional networks, develop new career pathways, or accelerate the path to leadership positions. For employers, mentoring programs have the potential to increase overall employee satisfaction, help to identify emerging talent, and either improve or further solidify an organization's culture.

Organizations, however, have not been able to tap into the true potential of what a well-thought-out, planned, and executed mentoring program can yield. Many formal mentoring programs within organizations unfortunately lack proper design, employ surface-level matching of mentors and mentees, or fail to give mentors and mentees the training and tools they need to have a successful relationship.

Here, we'll examine innovative mentoring programs at Coca-Cola and Zendesk—two companies whose programs have been critically lauded for their promotion of diversity, inclusiveness, ingenuity, and results-based outcomes.

Key Concept: Successful mentoring programs require in-depth planning and support from program leaders, training and expectations management for mentors and mentees, and firm commitment from all involved.

Coca-Cola's Journey to DEI Success

The Coca-Cola Company has been applauded for its efforts and commitment to DEI and mentoring. Among some of its accolades:

- The company has appeared on DiversityInc's Top 50 Companies for diversity 13 times since the program's inception in 2001.
- The company has received a 100 percent rating from the Human Rights Campaign based on its workplace policies and practices to support LGBTQ+ employees from 2008 and 2021.
- The company was cited for "Best Coaching and Mentoring Initiative" in 2020 by the Chartered Institute of Personnel and Development.

Coca-Cola's DEI ambitions, as detailed on its corporate website, are as follows:

- *Mirror Our Markets*—with a particular emphasis on having women hold 50 percent of senior leadership roles, and in the U.S., having the demographics of its workforce reflect national census data at all levels
- *Equity for All*—eliminate biases and inequities in all its policies and procedures
- *Inclusivity*—using its brand and image to "inspire and advocate for inclusion around the world"

Coca-Cola's road to recognition in the DEI space and as a model for other companies was not without difficulties and public scrutiny. Particularly, a historic class-action racial discrimination lawsuit from over 20 years ago would serve to be a major catalyst for Coca-Cola to reflect on its efforts (or lack thereof) on DEI and embark on a path toward sweeping change.

The Coca-Cola Company is, as of 2022, the third largest beverage company in the world, with $38.7B in revenue in 2021. The Fortune 500 company was founded in 1892 in Atlanta, Georgia. In addition to being the maker of the iconic beverage, Coca-Cola, it also owns the Minute Maid brand of juices and Dasani brand of purified water. The company distributes its products in over 200 countries via distributors, wholesalers, retailers, and an ecosystem of other partners.

The company has approximately 79,000 employees worldwide, with roughly 12 percent of its employees in the U.S. For 2021, women held nearly 40 percent of the U.S. senior leadership positions, with nearly 10 percent for Hispanics and nearly 10 percent for Blacks.

The Costs of Inaction and Not Listening to Employees

During the 1990s, the company was led by Douglas Ivester, its 10th chief executive officer. The company was on a very aggressive growth plan, which was already yielding favorable results. For 1995, the company's operating income grew 10 percent year over year, which was on top of an already 20 percent growth from 1994. Also in that period, it acquired the Barq's root beer brand, opened its first bottling operation in Vietnam, and even built the Coca-Cola Olympic City, an eight-acre park in downtown Atlanta to entertain spectators for the Centennial Olympic Games in 1996. On the surface, Coca-Cola was hitting the right points in terms of growth and revenue.

However, there had been growing discontent internally, chiefly among Coca-Cola's Black employees. Carl Ware, a Black senior vice president for the company at the time (and at that time, the only senior Black executive at the company), approached Ivester on creating a diverse committee to address the concerns and discontent among Black employees. Since joining Coca-Cola in 1974, Ware had a successful track record as a results-driven leader and an adept diplomat within the organization. When faced with a potential boycott of Coca-Cola products, led by the Rev. Jesse Jackson in the early 1980s due to its lack of hiring Black workers, Ware created a $50 million program to encourage partnerships with Black vendors.[1] Ware had been privy to many of the frustrations, biases, and aggressive behavior that Black employees had been subjected to for many years.

With the assistance of an external consultant, Ware, Ivester, and others—including Black employees and other senior leaders across the organization—held several discussions on the Black employee experience at the company. At the end of 1995, the external consultant concluded in a report to Ivester that the company repeatedly ignored or

---

[1] Encyclopedia.com, May 29, 2018. "Biography of Carl H. Ware."

discounted the experiences of Black employees, and that the company lacked a clear understanding of how diversity is a bona fide benefit to the organization. The report also included recommendations that could be implemented to promote more Black employees to senior leadership positions and cultivate an environment of tolerance for all diverse employees. Despite this, no further action took place at the conclusion of these committee meetings.

Fast forward to 1999—former Coca-Cola employee Linda Ingram and three additional employees filed a formal racial discrimination lawsuit against Coca-Cola, alleging that the company engaged in a pattern of discriminatory behavior against Black employees. They also charged that the acts of discrimination largely went unacknowledged and their offenders woefully under punished.

Ivester did not take the counsel of the external consultant's comments and suggestions or the opportunity to understand the background and the gravity of the allegations. Instead, Ivester chose to vehemently and publicly challenge the lawsuit. As the investigations continued, more allegations became known, and seriously harmed the company's public reputation. Ware resigned in 1999 (due to a management restructuring, he had been effectively demoted), and perhaps further fueled the anger and animosity among many Black employees.

Positivity from Turmoil

The fallout of the 1999 lawsuit, while not immediate, did lead to positive changes for the company. By 2000, Ivester was let go, and the company reached a settlement. In addition to a monetary settlement of $192M to be paid to the plaintiffs (current and former Black employees), the settlement also mandated the creation of an external task force to implement programs to retain and promote employees of color, at an additional cost of $36M ($62.3M in 2022 dollars), and with every area of the company analyzed from the executive level on down. For a five-year period, the external task force evaluated the changes that the company made that they considered progress and areas for improvement and made their findings publicly available through a series of four reports.

The Problem: How to rebuild trust and establish a more equitable work environment through mentoring, following a time of crisis.

In 2001, as part of the task force's recommendations, the company piloted a new mentoring program to provide, as the company described, "all employees a way to share experiences and expertise that will result in professional development and personal growth." Mentoring was specifically designated by the task force as one of nine HR systems that the company should focus on.

The pilot program had sponsorship from senior leadership and was communicated broadly to U.S. staff. For the pilot, 100 pairings of mentors and mentees were formed to create a formal mentorship program. The program architects addressed the shortcomings of other mentoring programs by innovating the following features:

- A formal application and matching process. Mentors and mentees had to self-identify as wanting to participate in the program. Mentees, who company's leadership felt may have disproportionally not had access to senior leadership or to career development activities were encouraged to participate. Mentors were assessed on their ability to coach and lead, in addition to their ability to offer wisdom and expertise. Great care was put into matching mentors and mentees beyond high-level interest match, such as communication styles, conflict resolution styles, and personalities.

- Training. Both mentors and mentees received training at the start of the program and on an ongoing basis. The training focused on what it meant to be a "mentor" and "mentee," as well as the responsibilities each had individually and to each other.

- A formal agreement. Each mentoring pair would create an agreement that outlined their development plan during the duration of the program, the frequency of meetings, as well as the confidentiality of items that may come up during mentor and mentee meetings.

- Support. Mentoring coordinators would have touch points with each of the pairs. They would act as resources for the mentees and mentors for any questions or concerns they may have.

- Options. For those who did not want to participate in the formal program, or couldn't, a self-study guide was created, and group mentoring sessions were offered.

- Measurable and actionable feedback and data. Participants had multiple opportunities throughout the program to offer their thoughts, through questionnaires and focus groups, on whether the program was beneficial to them. Program coordinators also measured the program's participation by demographics.

Measurable Results

In 2002, in the task force's first official report, Coca-Cola was praised on how well the pilot program had been designed and implemented, even commenting that their efforts exceeded what had been outlined in the settlement agreement.

According to the feedback from employee engagement surveys, many participants found their experience positive and valuable, with no significant differences in satisfaction along racial or gender demographics. Due to the positive feedback, the pilot was used as a model for additional mentoring programs across business divisions and departments, scheduled to be rolled out in late 2002 and 2003. The company also planned to weave in the use of formal mentoring programs as part of a broader, more integrated employee career development plan.

An area of concern for the task force was the advancement of women and people of color in more senior-level positions in the company. While advancement and promotion of women and people of color was contingent on the effectiveness of the eight other systems (e.g., "Career Planning," "Succession Planning"), the task force felt that the mentoring program could help aid these efforts.

Over the course of the next decade, the company continued to adjust and refine the program based on participant feedback and task force recommendations. Executive-level employees began to participate as mentors. In 2004, it launched a "Networking for Success" program, where company leaders would speak and interact with other employees. By 2005, the end of the initiative, the results of the program were measurable:

- More than two-thirds of the mentees who participated in the mentoring program remained in the company. Of that figure, 45 percent of them were employees of color. In contrast, Coca-Cola lost 20 percent of its Black employees, 5 percent of its Hispanic employees, and nearly 30 percent of its female employees year over year from 2000–2002.[2]

- The remaining employees advanced their careers within the organization—42 percent received promotions, while 38 percent made lateral moves across the organization.[3]

---

[2] First Annual Report of the Task Force, United States District Court, Northern District Court of Georgia; Ingram, et al. v. The Coca-Cola Company, 2002.

[3] Fifth Annual Report of the Task Force, United States District Court, Northern District Court of Georgia; Ingram, et al. v. The Coca-Cola Company, 2006.

- In 2010, 15 percent of executive roles in the United States were occupied by Black employees. This is up from almost 2 percent since the original lawsuit was filed in 1998.[4]

Moving on from the Past

Coca-Cola's adoption and embrace of high-quality, outcomes-based mentoring methods have certainly paid off, and it is evident that it plays a core part of the company's DEI efforts. Mentoring programs aimed at women and people of color, from early career to senior leadership, can be found across departments and functions. The turnaround and success stories that the mentoring program serves as a model for other businesses to incorporate mentoring programs as a means to cultivate and promote underrepresented talent.

In more recent years, Coca-Cola has experienced employee turnover, and more pointedly, sharp attrition of Black leaders in their executive ranks. While it is unclear of the root cause(s), they were also likely not immune to the effects of the Great Resignation (we touch more on this and employee retention in Chapter 8, "Rethinking Retention Through the Lens of DEI"). Put simply by Valerie Love, senior vice president of human resources, "We didn't keep our eye on the North Star."[5] Beginning in 2023, they intend to reignite their efforts to improve recruiting and retaining Black employees and executives. While no organization's leadership team wants to address past mistakes, discrimination, and inequities in such a public (and costly) fashion, it does take a significant amount of courage to face them head on. Recognizing what was at stake, Coca-Cola took responsibility and collaborated internally and externally to address the problems in a meaningful way.

It can be argued that had the racial discrimination lawsuit not become known and the public scrutiny along with it, Coca-Cola may have never been prompted to embark on these changes. Yet, it could also be argued that the lawsuit and settlement themselves would not have necessarily been the catalysts for profound change either. We've all likely seen companies that were slapped with huge fines for all types of malfeasance and then continued to engage in the very behavior that they were punished for. The change that took place at Coca-Cola could not have been successful on just a desire for compliance alone. It required a committed cultural shift, and a strong leadership team with the courage to act, and in so doing, to confront the legacy of Fixed Attitudes and Fixed Practices long held at the company.

[4, 5] *Wall Street Journal*, December 16, 2020. "Coke's Elusive Goal: Boosting Its Black Employees."

THE RESEARCH BEHIND IMPACTFUL MENTORSHIP PROGRAMS

Cultivating mentorship programs that will serve and promote diverse populations takes more than just good intentions. Careful program design is key to not only reaching these audiences, but to ensure that the mentoring experience is meaningful to mentees.

To that end, The EDGE in Tech Initiative has sought to help companies in creating effective programs. The mission of EDGE (Expanding Diversity and Gender Equity), per director Jill Finlayson, is to promote the "participation, persistence, and advancement" of women and people of color within science, technology, engineering, and mathematics (STEM) professions. This work is done through research projects, events, and leadership roundtables. The initiative's work has spanned four of the University of California's (UC) campuses—Berkeley, Davis, Santa Cruz, and Merced.

The initiative seeks to promote diversity and inclusion through the lens of innovation and entrepreneurship at Berkeley and throughout the UC system. As Finlayson explains, this is not only looking at best practices, but where improvements can be made to better serve their communities. Mentorship specifically came into focus as there was a sense that target populations were not participating or actively engaging in these programs.

Anita Balaraman, the Principal Investigator on the research representing UC Berkeley, along with her colleagues at UC Davis and UC Merced began to examine who was (and wasn't) participating in the programs. They also began to look at the impact of networks. According to Balaraman, a key benefit of mentorship programs is to increase the awareness, access, and acceptance into a chosen field of study or career, and be able to thrive and persist in the respective environments. Research indicates that access to social capital via mentoring is critical for historically excluded students' sense of belonging, self efficacy and retention. However, as Balaraman points out, "most mentoring programs today focus on skill acquisition to the exclusion of expanding participants' self-efficacy and social capital." What impact, then, can a mentor have on a mentee that either has no network, or an uneven network? In what ways can shared values and power dynamics between mentors and mentees affect the overall efficacy of these programs? Where can there be intervention to improve outcomes?

Per Balaraman, "the aim of the study was to identify characteristics of mentoring programs that benefit (or do not benefit) women, BIPOC, and first-generation college students and increase their retention and continuation in STEM. The hypothesis was that shared values and power dynamics can drive the success (or failure) of mentoring these students in STEM. Specifically, we studied the impact of patented technology 'Epixego'—an online mentoring and employment ecosystem—and the accompanying training program that both explicitly incorporate shared values and account for power dynamics in mentoring. The research was an intentional collaboration across UC Davis, UC Merced, and UC Berkeley, with the former two having the distinction of being Hispanic-Serving Institutions (HSI - a qualifying educational institution that has at least 25 percent Hispanic students enrolled) in a near-peer mentoring model." In this model, a mentor is typically a few years or life stages ahead of their mentee. Research indicates that access to social capital via mentoring is critical for historically excluded students' sense of belonging, self-efficacy, and retention.[6] The research used a mixed-method approach consisting of a quantitative assessment of the mentoring intervention using pre- and post-intervention surveys and qualitative data from focus groups."

Balaraman continues—"Existing mentorship measures were adapted and integrated from social influence theory, social cognitive career theory, social network theory, and Global Measure of Mentorship Practices.[7] The measures corresponding to a mentor or mentee's past mentoring experience were calibrated against the post-intervention scores. Shared values between mentor and near-peer mentee increased by 27.03 percent, and mentee's STEM self-efficacy (measured via occupational identity, & social capital) improved by 23.19 percent and 35.15 percent respectively."

Balaraman concluded, "the study contributes to research on the pedagogy of effective mentoring programs and interventions. Assuming mentees and mentors have the skills and tools needed to develop a successful mentoring relationship disadvantages mentees who lack sufficient social capital to connect to their mentees.[8,9] While some progress has been made in educating mentors and mentors,[10,11] this study advances frameworks, tools and metrics to implement and measure mentoring relationships.

Mentoring Innovation at Zendesk

Let's now take a look at mentoring at tech company Zendesk. By all accounts, Zendesk is a company that has created DEI and talent development programs with strong results. Yet in spite of its successes, leadership from one of Zendesk's employee resource groups found a

[6] Holloway-Friesen, H. (2019). The Role of Mentoring on Hispanic Graduate Students' Sense of Belonging and Academic Self-Efficacy. Journal of Hispanic Higher Education, 1–13. https://doi.org/10.1177/1538192718823716

[7] Ash, R. A., & Dreher, G. F. (1990). A Comparative Study of Mentoring Among Men and Women in Managerial, Professional, and Technical Positions. Journal of Applied Psychology, 75(5).

[8] Pfund, C., Byars-Winston, A., Branchaw, J., Hurtado, S., & Eagan, K. (2016). Defining attributes and metrics of effective research mentoring relationships. AIDS and Behavior, 20(S2), 238–248. https://doi.org/10.1007/s10461-016-1384-z

[9] Pfund, C., House, S., Asquith, P., Spencer, K., Sillet, K., & Sorkness, C. (2012). Mentor training for clinical and translational researchers. W H Freeman & Co.

[10] Gandhi, M., Johnson, M. Creating More Effective Mentors: Mentoring the Mentor. AIDS Behav 20 (Suppl 2), 294–303 (2016). https://doi.org/10.1007/s10461-016-1364-3

[11] Pfund, C., Handelsman, J., Branchaw, J., & Branchaw, J. (2014). Entering mentoring. W.H. Freeman.

potential area to improve upon its already successful mentoring programs and help bring more women to leadership roles in the process.

Headquartered in San Francisco, California, and founded in Copenhagen, Denmark, in 2007, Zendesk is a leading provider of cloud-based customer service and relationship software. Zendesk achieved $1.3B in revenue in 2021 and boasts over 100,000 clients in more than 160 countries. Zendesk's services are used by major companies like ride-sharing app Uber, e-commerce platform Spotify, automobile manufacturer The Ford Motor Company, and British multinational retailer Tesco, to name a few.

The Problem: Representation of women in senior leadership positions lagged behind their male counterparts, and the goals of Zendesk's leaders.

In 2021, the company had 5,860 employees worldwide, and places a premium on cultivating a diverse and inclusive culture that fosters belonging. "We're being very deliberate about building a diverse and creative company where all employees feel empowered and comfortable being their authentic selves," says Mikkel Svane, the company's chief executive officer. Their efforts are paying off—the company has been recognized by the Great Place to Work Institute in 2017 and 2018, and as a "Best Place to Work" winner by company review website Glassdoor.

The leaders at Zendesk recognize that a continued, sustained commitment is necessary to ensure that their inclusive culture remains strong. They also recognize, much like Citi, not to rest on past successes and to continue to improve where they can.

Take, for example, the representation of women employees who hold leadership positions in the company. At the end of 2018, women held 32 percent of the senior leadership roles, including those that were director level and above, but Zendesk leadership was far from satisfied with that number and identified mentorship as one strategy to accelerate the professional growth of its female staff

Now, Zendesk has several mentorship programs available to its female employees. Since 2014, the company has partnered with Everwise (a digital platform used to deliver mentoring and coaching programs; now owned by Torch) to create a mentoring program aimed at high-performing/high-potential female employees who were looking to advance their careers in the organization. The program has been very successful by all measures, but there were also some limitations—by meeting virtually, mentors and mentees didn't have the opportunity to meet in person on a regular basis. Because the platform was developed by a third party, it may not always capture the culture and nuances that a home-grown, internally created program would create.

Zendesk's Women Mentorship Program: Initial Pilot Program

Zendesk sponsors and supports several internal employee resource groups to support its diverse staff. Among them is the Women of Zendesk ERG, charged with helping to support and advance female employees within the organization. Beginning in 2018, the ERG is supported by several female leaders in the company, including at the executive level.[12]

Over the course of several conversations with the leaders in the ERG, a common thread emerged—many members acknowledged the positive impact that mentoring (whether formal or informal) had on their career trajectory, both in and outside of Zendesk, and wanted to provide a support mechanism for women to advance within the organization. Taken further, the ERG saw mentorship as a means of creating a supportive network for employees and continues to promote a diverse and inclusive environment. Gomez, a Latina whose parents emigrated from Mexico and El Salvador, respectively, emphasized the difficulty in implementing diversity within an organization." It's very hard to do. It takes years and a lot of work and a lot of focus. And so I really felt passionate there to to be the role model and really do what I could to bring more diversity to ZenDesk and to really, not just Zendesk but just the corporate world in general."[13] From these conversations, the ERG developed a pilot mentorship program, led by Zendesk's former chief financial officer Elena Gomez.

The ERG's initial pilot program was conducted for three months and in two separate locations—their headquarters and their Madison, Wisconsin office. Their pilot program included the following features and attributes.

Application and Matching Process

Although there were many who expressed a desire to be mentored (more than 200 women applied to be mentees), only 50 women were available as mentors. As this was intended to be a one-to-one mentoring model, only 50 women could participate in the pilot as mentees.

---

[12]McElhaney, K., Mancia, A. C., & Rustagi, I. Zendesk: Building female leaders through mentorship. Copyright 2019, by The Regents of the University of California. All rights reserved. Do not copy, reproduce, or otherwise disseminate without written permission from the Berkeley-Haas Case Series. For additional permission or reprints, contact: cases@haas.berkeley.edu.

[13]Elena Gomez, BS 91—passionate about diversity and inclusion in the Workplace. Haas Podcasts. (2022, October 15), from https://haaspodcasts.org/podcast/elena-gomez-bs-91-passionate-about-diversity-and-inclusion-in-the-workplace

To reduce the introduction of bias in the selection process, the pilot administrators used a randomized spreadsheet to select the mentee participants. Additionally, pilot administrators also carefully made mentor/mentee matches manually, based off of application responses and a focus on cross-departmental pairings, to encourage interaction and communication with women who may not already have much interaction with each other.

Support

In addition to the biweekly mentor and mentee meetings, the pilot introduced "Community Sessions." Here, mentees and mentors would meet with other pairs to further their discussions on career and personal development, as well as for mentors and mentees to have an avenue to continue to build out their professional network.

Measurements for Success and Feedback

Mentors and mentees were asked to provide feedback via surveys on a regular basis to understand what they enjoyed about the program, where improvements could be made, and where numerical scores would be given. The outcomes of the pilot would be presented at an employee town hall to ensure transparency and accountability for all involved in the program.

Pilot Observations: What Worked and What Didn't

At the conclusion of the pilot, the program team evaluated feedback that they received from participants, as well as reviewed their own work during the process. From a positive standpoint, they found that many participants felt that their participation was beneficial, in that they felt they received support from Zendesk in their career and personal development, and that they now had a richer, broader network to reach out to. Participants found the "Community Sessions" incredibly helpful, solidifying the sense of community among participants. The program had also received positive acclaim throughout the organization.

The feedback also uncovered some areas of improvement. Participants felt that the short time frame didn't allow for the deepening of relationships. Plus, for employees with particularly busy schedules, mentor and mentee meetings may have been shortened or canceled altogether, minimizing the impact that mentoring may have. While there was a significant amount of time devoted to pairing, and they were provided handbooks and instructions on how to structure their one-on-one meetings, mentees and mentors hadn't had the benefit of a formal training period, causing both to be at a loss for conversation topics during their one-on-ones and a bit unsure of what they should expect from one another.

Improving on Success

The pilot program team felt very empowered from the success of the pilot and wanted to build upon their initial successes, while making the program accessible to more women, with more impactful relationships. To help in this effort, the pilot program team partnered with the Center for Equity, Gender and Leadership, part of the University of California Berkeley Haas School of Business, and Berkeley Women in Business to understand how they could improve the program.

Together, they produced a rather innovative approach—they reached out to UC Berkeley-Haas students to participate in a student competition, which was held in the latter part of 2018. During this challenge, the students were asked to offer solutions to the problems that surfaced during the pilot—specifically around scaling the program across the company, proper preparation for mentees and mentors, and what metrics should be measured to ensure the program's efficacy. A total of 26 teams comprising over 150 students presented their pitches to judges.[14,15]

Many ideas were discussed and considered, but the winning team's idea centered around the development of a smartphone application that would allow mentoring program participants to interact with one another anytime, anywhere.

Measurable Results

By the end of 2019, with subsequent and robust improvements in execution, the program was adjusted to be implemented in several of Zendesk's offices, both in the U.S. and in its international offices. The program would serve as an inspiring model for the mentoring programs Zendesk would later develop, including the Ignite Development Program, whose focus is on advancing employees from underrepresented demographics.[16]

Today, nearly 44 percent of leadership roles (director level and above) are held by women at Zendesk, up from 32 percent in 2018. This compares to just over 25 percent in the tech sector generally, according to a 2022 study by Deloitte.

[14]Zendesk. Berkeley Haas. (2021, February 10), from `https://haas.berkeley.edu/corporate-partners/our-partners/zendesk`

[15]Mancia, A. (2019, February 13). Zendesk Case competition. Medium, from `https://berkeleyequity.medium.com/zendesk-case-competition-4e19d40f862a`

[16]Torch. (2022, July 15). Zendesk Fuels Inclusion and career growth through leadership development. Torch, from `https://torch.io/blog/zendesk-fuels-inclusion-and-career-growth-through-leadership-development`

Let's examine how both Coca-Cola's and Zendesk's approach to designing and implementing mentoring programs fits in line with innovation principles (Table 5.1). Table 5.2 looks at the innovation threats and actions that could prevent them.

Table 5.1: Aligning Mentoring Programs with Innovation Principles

PRINCIPLE NECESSARY FOR A CULTURE OF INNOVATION	PRINCIPLE IN ACTION
Courage	■ Coca-Cola: Learning from past mistakes and transgressions and using those learnings to make sweeping and necessary changes across the company.
Risk-Taking	■ Coca-Cola: Committed a sizeable number of financial and other resources to implement its mentoring program and improve other HR systems, without the assurance that its efforts would be successful. ■ Zendesk: Employed creative means and diverse audiences to make improvements to its already successful mentoring programs.
Trust	■ Coca-Cola: Leaders needed to trust that the recommendations and the feedback of the appointed settlement task force were in the best interests of the company, as well as one another to carry out the monumental changes that were being asked of them.
Collaboration	■ Coca-Cola: Collaborated with HR, DEI, and talent development leaders (in and outside the company) to understand where weaknesses in their mentoring program persisted, and how they could be improved. ■ Zendesk: Leveraged leadership in various parts of the organization's structure, as well as the expertise of experts external to the organization, to design and implement a successful mentoring program.
Leadership	■ Zendesk: Being creative and taking the initiative to further innovate an already successful mentoring program, and subsequently asking for resources and sponsorship to do so.

Table 5.2: Innovation Threats and Preventative Actions

INNOVATION THREATS	ACTIONS TO AVOID THREAT
Arrogance	▪ Coca-Cola: The senior leadership at the time of the Ingram discrimination lawsuit put the company in a precarious position. Rather than taking this as an opportunity to learn about and acknowledge short-comings, they challenged the findings at the cost of the company's reputation. Realizing the irreparable harm that was caused, he was asked to resign so that the necessary work of rebuilding the company's reputation and healing its relationship with employees of color could begin.

Conclusion

While the genesis of the mentoring programs for Coca-Cola and Zendesk were vastly different, we have seen that the success of both was contingent on several shared factors. The desire and capacity to help others is certainly beneficial but is not enough on its own. Mentees can, and should, be held accountable for being active participants in formal programs. But it is on the shoulders of an organization's leaders to take care in finding mentors with nurturing mindsets, provide ongoing training and support for participants, and offer the means to provide continuous feedback so that the program can improve and grow.

Summary

- Mentoring can be a wonderful resource to help advance diverse talent and can be mutually beneficial to employers and employees alike.

- Like any other workforce or talent development program, a successful mentoring program requires considerable time dedicated to the design and implementation alone.

- Care should be exercised when making mentor and mentee matchups; while matching a mentor's skills and experience to a mentee's interests and goals is necessary, understanding both the mentor's and mentee's communication styles and other personal traits is just as important for mentoring to be impactful.

- Providing training to both mentors and mentees helps level set expectations on realistic outcomes, as well as helps both partici-pants understand one another better.

- Even the most successful mentoring programs require a mechanism for participants to provide continuous feedback to help improve the program for future participants. It is this act of constant fine-turning and focus that can lead to more substantial innovation in programs.

Looking Beyond Traditional Talent Sources for "Hard to Find" Roles

In recent years we have seen a trend of companies reconsidering the need for college degrees across a number of entry-level roles. As the labor market tightened, hiring managers reevaluated job descriptions to assess skills most relevant to performing required tasks. Waiting to find "unicorns" who met countless requirements—many "nice to have" but not essential—meant jobs remained opened while existing staff risked burn-out after picking up the slack.

This chapter investigates how visionaries at two leading companies—Northrop Grumman and Tessco—took the idea of hiring talent with non-traditional backgrounds a step further. Each innovated new systems for identifying, training, and career-pathing high-potential, diverse talent from unexpected sources. Each has experienced success beyond what they and their colleagues thought possible.

Key Concept: Identifying and hiring candidates based on skills and potential, rather than degrees, can unlock entirely new sources of exceptionally creative and highly motivated talent. Doing so often requires a level of risk-taking, trust, and leadership to build systems that supplement the traditional college recruitment and hiring process.

Northrop Grumman and Tessco: Shifting Long-Standing Perceptions of Who Can Succeed

In this chapter we'll explore how leaders at two companies in very different industry sectors successfully challenged and changed systems to enable individuals without bachelor's degrees or years of related experience to thrive in hard-to-fill roles. Each company arrived at the decision to recruit non-traditional talent for different reasons but shared a common goal to build career onramps and developmental pathways for individuals in low-income Baltimore neighborhoods who were too often shut out from well-paid corporate jobs.

Context on the economic and demographic characteristics of Baltimore will be helpful for this discussion. Baltimore ranks 30th among major U.S. cities for population, just behind Louisville, KY, and ahead of Milwaukee, WI. The regional economy is anchored by one of the 10 largest ports in the U.S., and by world-class educational and healthcare institutions, including Johns Hopkins, University of Maryland, Morgan State, MedStar Health, and Mercy Health Services. Federal and state agencies, including the Social Security Administration, are significant employers representing about 20 percent of all regional jobs. With proximity to numerous high-growth biotech companies and defense contractors, jobs in life sciences and technology—especially cybersecurity—have exploded in recent years. According to the Maryland Department of Labor, Baltimore has the third highest density of cybersecurity jobs in the nation.

Yet, access to good jobs in Baltimore's dynamic economy has been elusive for many Baltimore residents. Unemployment rates in the lower income east and west sides of Baltimore City range between 16 and 21 percent, compared to regional unemployment rates of 3.6 percent in May 2022. Educational achievement follows a similar pattern. Adults with a bachelor's or advanced degrees comprise roughly 33 percent of Baltimore City's population according to Baltimore Neighborhood Indicators Alliance, which is on par with the U.S. average. However, of those living in Baltimore's underserved and low-income neighborhoods, only about 15 percent have bachelor's or advanced degrees.

A long history of discriminatory housing, transportation, and education practices and policies contributed to the economic marginalization of many Baltimore neighborhoods. Public schools in the region have underperformed for decades with classes taking place in buildings with insufficient infrastructure, some lacking proper heating and air conditioning. Transportation initiatives that physically segregated many of

Baltimore's Black communities further entrenched cycles of generational poverty. With few public transit options, those living in the west and east sides of the city long faced difficulty getting to jobs in downtown Baltimore or in the outlying suburbs.

Breaking cycles and forging new paths require a degree of systems-level thinking and courage. Common to both case studies featured in this chapter are determined leaders keen to spark creative output and performance. At the same time, both sets of leaders wanted to take some responsibility for helping to improve the economic outcomes of residents in the communities where their businesses operated and were prepared to tackle systems change within their own companies. In the process, they both managed to break through Fixed Attitudes and Fixed Practices to build wholly different models of talent development. While the new talent practices reviewed here won't solve the problems of income and social inequities in Baltimore, they do offer innovations that have already changed the economic trajectory of the staff members hired and pave the way for many others. They have also altered the way each company recruits and develops talent, with a greater focus on inclusion.

We'll start with the case of Northrop Grumman.

Northrop Grumman: Focus on Novel Thinking and New Talent

Northrop Grumman is a $30 billion aerospace, defense, and security company with the majority of its business providing services to the U.S. government, principally the Department of Defense and intelligence community. With over 90,000 employees worldwide, Northrop Grumman has a long history of pioneering technologies in areas as diverse as autonomous vehicles, cybersecurity, advanced aircraft, logistics for U.S. emergency services, communications satellites, and space systems. The company attracts scientists and engineers from top universities across the U.S. who often work in a top-secret, classified environment.

In 2022 Northrop Grumman ranked 20th among DiversityInc's Top 50 Companies committed to diversity, and second in the aerospace and defense industry.

The Problem: Standing up an experimental cybersecurity lab for a major defense client with talent who could think "without boxes," on a tight budget, and with an intrapreneurial mandate that required operating at the margins of the company's traditional practices.

Andrew Parlock was hired by Northrop Grumman early in 2017 to revitalize a flagging cybersecurity project for a significant customer. He accepted the offer fully understanding that success would require fortitude, ingenuity, resilience, and a whole new way of thinking. It was exactly the kind of challenge he wanted after a long career at the intersection of business development, engineering, and national security.

Within his first few months on the job, Andrew identified an opportunity he believed would be essential to rebuilding the client partnership: standing up a research and development lab to test a range of new cybersecurity technologies to support the client's objectives. He proposed that the new lab could transfer innovative technology developed in an unclassified environment to a classified category benefitting national security. He called it the "Emerging Technology Innovation Lab," or ETIL. Andrew's proposal was approved by his supervisors, although it was provisioned with a small budget since the lab cost would be incremental to the original scope of the contract.

Undaunted, Andrew set about staffing up the new lab. His first hire was Austin Cole, an engineer and solutions architect from the Department of Defense with expertise in end-point security—Austin knew what it took to secure devices like mobile phones and laptops from malicious actors. Andrew added to the team a second seasoned cybersecurity professional with nearly 30 years of systems engineering experience. To round out the team and stay within budget, Andrew hoped to bring on a couple of recent college graduates, but the recruiting season had effectively closed for the engineering talent he wanted to land.

Andrew was inspired to take a completely different direction after a meeting with the Baltimore mayor's office about skilling up young adults from Baltimore's most underserved neighborhoods for the surging number of local cybersecurity jobs. He was struck by one statistic cited: if a young person from conditions of poverty doesn't have a career by age 27, they will likely be stuck on the "merry-go-round" of minimum wage jobs their entire working life. As a resident of Baltimore City and a frequent mentor of young people not unlike those cited in the study, he realized he had an opportunity to make a difference—and just maybe address his staffing needs.

Out of that meeting with the mayor, he was introduced to the team at NPower Baltimore, which was focused on training young adults in a range of tech skills and connecting them to career-track jobs. He figured that a few well-selected NPower students could help set up the lab and assist with basic-level tasks, freeing up the seasoned engineers to focus on more complex problems. As luck would have it, NPower's last cohort of students was just completing their training and certifications and was available for internships. Andrew enlisted Austin's support to figure out how to recruit and integrate talent without the background typical of most Northrop Grumman engineering hires.

Roadblocks and Pathways

What seemed like a straightforward talent solution was quickly stymied by long-standing HR practices at the company. The first obstacle was the format and intention of Northrop Grumman's internship program. "Interns" were narrowly defined as college students majoring in science, technology, engineering, or math (STEM) and selected for summer employment months earlier. The formal internship program proved successful over the years as a pipeline for full-time talent, but it was reserved only for college students. So, the young people Austin hoped to bring on could not be called "interns."

Nor could they be full-time staff members. The second obstacle was the company's practice of requiring a minimum four-year college degree. The majority of the participants in the NPower program lacked a college degree, although several had one or two years before dropping out, often for lack of financial resources.

Austin put his engineering mindset to work and—with the support of his HR partners—found a creative solution by hiring the young talent as sub-contractors (referred to as "trainees" in this case study). NPower formally served as the vendor supplier, identifying strong candidates for Austin's project and handling payroll. With the size of the contract, the approval could be made at the program level, which meant Austin could sign off on the arrangement.

Both Andrew and Austin believed that if the new talent brought on as sub-contractors performed well, there was a possibility they could be converted to full-time hires once the lab's assignment was complete. Andrew and Austin knew they would need to take some professional risks by challenging the present employment rules, but also believed very deeply in the contributions of diverse backgrounds to problem-solving. They also shared a belief in the need to build pathways for exceptional talent held back by barriers of socioeconomic conditions and stereotypes.

NATIONAL INITIATIVES TO ACCELERATE TRENDS IN SKILLS-BASED HIRING

In March 2020 the National Bureau of Economic Research released a working paper (Peter Q. Blair, 2020) exploring the characteristics of some 70 million working adults across America without bachelor's degrees. The authors coined the term "STARs" (Skilled Through Alternative Routes) to describe those who have work experience and skills that could position them for transitions to higher-wage jobs.

The study revealed striking statistics on the gap between demand and supply for college-educated workers. The Bureau of Labor Statistics reports that from 2007 to 2016, 74 percent of new jobs created required a 4-year college degree, double the share in jobs created prior to 2007. If this trend continues, only 26 percent of new jobs will be available to the 58–62 percent of U.S. workers without a college degree. From the employer perspective, only 38–42 percent of workers will be eligible for 74 percent of new jobs. Serious challenges will confront employers in the near term, along with worse economic outcomes for workers with lower levels of formal education.

Since the release of the findings, national organizations such as Opportunity@Work and OneTen have partnered with labor networks and training providers to help individuals map the skill content of their current occupations to the skill requirements of higher-paid positions. Similarly, these organizations advise companies on how to expand their recruitment nets to reach individuals in skill-adjacent jobs who, with some training and upskilling, could readily succeed in in-demand jobs.

Shared Values Shaped by Common Experiences—and a "Secret Mission"

As the lab got up and running, the goal of converting the subcontractors to full-time employment became what Andrew described as the "secret mission." As we have seen in other case studies, the seeds of organizational change are often planted by a handful of passionate leaders driven by their own personal experiences and who are willing to take risks.

In the case of Andrew and Austin, their respective journeys were remarkably similar, although separated by about two decades. Both men were the first in their families to graduate from college and both spoke of their blue-collar, working-class roots. They are both sons of fathers who pursued similar paths; Austin's dad was an electrician and Andrew's was a mechanical technician.

What stood out in our separate conversations with each of them was their admiration for their fathers. Each took pride in the careers their dads had built, but also revealed they believed their fathers could have been engineers or computer scientists had circumstances been different.

Austin shared the story of how his father helped him find a path to college, which was also his first experience with the concept of "networking." His dad was employed by a small government electrical contractor and was frequently on site at a federal agency where he became friendly with its staff members. One day he spied a flyer posted on a bulletin board announcing a high school internship program at the Department of Defense for those interested in engineering. A few inquiries and phone calls later, Austin's dad secured an interview for his then 15-year-old son that led to a summer job, and later helped Austin earn a scholarship to the University of Maryland.

Austin was over-the-moon about the internship. He describes his high school as one that lacked tech resources and had few STEM courses, but Austin was intrinsically motivated and made the most of his summer job. He later reflected on the respect he has for those who come from environments with fewer resources than he had, but still manage to compete for top opportunities. "Some young people have access to bootcamps and loads of after-school resources, which give them a head-start. Those who succeed despite limited resources make an impression—they can often creatively overcome all sorts of challenges in the workplace."

Selecting the Right Talent

The first young person subcontracted from the NPower program didn't work out, which presented a setback to Andrew and Austin's plans to identify talented individuals from non-traditional backgrounds for roles in the lab. Naysayers who questioned their talent plan were quick to enumerate why others without college degrees wouldn't work out either. Austin's colleagues who felt this way often cited that college graduates were generally more polished, had better communications skills, and were good "cultural fits."

The individual first hired was keenly interested in—and good at—hands-on IT work. In fact, he has gone on to have a fine career in IT installation and support with another company. Austin realized the poor fit was more about zeroing in on the specific skills he needed in the lab, and less about the aptitude of their first hire. The role required *both* hands-on IT hardware skills as well as an engineering orientation with abstract problem-solving capabilities.

In addition to working with NPower's career placement team, Austin decided to speak directly with the senior instructor who could expertly assess the problem solving, analytical, and logic skills of those in the class with the potential to succeed as software engineers with further training.

With a more refined set of requirements, the NPower team identified new candidates they believed could succeed. Austin selected three. None had a college degree. These three individuals thrived, in large part because of the on-site training and mentoring Austin and Andrew provided. Astrid Portner, an additional experienced engineer hired later to join the team, also proved to be a key partner in supporting the success of the candidates.

Onboarding and Upskilling: "Building Software Engineers"

"Even the brightest college grad with an engineering degree is not ready to be productive on Day 1, or Day 30," observed Austin. He recognized that even those hired through Northrop Grumman's traditional channels required role-specific training and benefitted from the company's formal onboarding program. Because the individuals hired for the start-up lab were subcontractors, Austin and Andrew formulated their own plan to successfully transition the three new hires. The hires where subcontracted for 16 weeks. By week 12, they would be evaluated and either informed of a contract extension or given 4 weeks to wrap up their projects.

Austin willingly took on the charge to create a ramp-up curriculum and on-boarding map to give the young trainees the best possible opportunity for success by week 12. What emerged was a comprehensive experience that included significant training, mentorship, and a profound camaraderie forged in a scrappy start-up environment where resilience and ingenuity were valued, independent of academic credentials.

In describing his curriculum focused on software development skills, Austin notes, "I basically compressed two semesters of computer science in ten weeks, building on the foundational coding introduction the subcontractors got through the NPower program." He broke down the first 12 weeks of the on-boarding, training, and work experience as follows:
Weeks 1 and 2

- Fully focused on training and orientation.
- Learned about Northrop Grumman products, history, and culture.
- Introduced to the basic building blocks of software skills needed to complete their assignments in the lab with a heavy emphasis on Python skills.

Weeks 3 through 6

- Intensive coding training and a deep dive into the projects they would be assigned.
- Shadowed a senior engineer 25 percent of the time.
- Began to attend daily "stand-up" meetings to hear what the other engineers did and the progress they were making against goals.
- Assigned supplemental work to build both professional and software skills (this component would extend through the entire duration of the 16-week period).

Weeks 6 through 10

- Contributed to active projects, including completing coding assignments overseen by a senior engineer.
- Spoke about their own work during morning stand-up meetings, developing presentation skills at the same time.
- Tapped to participate in impromptu problem-solving sessions with the senior engineers.
- Offered opportunities to interact directly with the client.

Weeks 10 through 12

- Independently contributed to assigned projects.
- Collaborated with colleagues across the lab on evaluating and addressing a range of real-time cybersecurity concerns.

Throughout this period, the trainees had a couple of hours on "Learning Fridays" to dive deeply into a new technology of interest to them, or to pursue a passion project. Austin thought about his own professional experiences observing, "Training is a hot button issue for engineers. They value employers who invest in their access to online training, who allow up to 10 percent of their time to work on something innovative, or who send them to conferences to hear about new research." He elaborated, "Too often training is viewed as a retention strategy during a tight labor market but should be an ongoing investment independent of the economy." He went on to add that in his experience across different organizations, many engineers are willing to take a slightly lower salary if their employer consistently invested in their growth and development as technical professionals. Austin concluded, "And engineers with this orientation toward continual development and self-improvement are generally the best employees."

"Relentless Focus on Culture"

Working with the trainees to upgrade their technical knowledge was central to the onboarding experience. They all had basic-level CompTIA or other industry-recognized certifications before joining Northrop Grumman as subcontractors but lacked the more robust software development skills the lab needed. But to Andrew's thinking, what was even more important was what he described as a "relentless focus on culture." Andrew believed that the right culture was necessary not only for the trainees to succeed, but essential for all members of the team to thrive, for innovation to flourish, and for the customer relationship to succeed. Austin shared Andrew's interest in creating an environment where colleagues were supportive of the new trainees and one another, where there was room to fail fast and learn quickly, and where frequent customer engagement was a priority.

Building the right culture required an intentional focus on both the physical environment and expected behaviors, attitudes, and activities, such as:

- **Lab Space:** With its shoestring budget, Andrew and Austin found empty storage space away from the main office building to set up a space that looked and felt different from any other work environment at the company. Describing a setup more akin to a Silicon Valley start-up, Andrew recalls, "We had white boards and lab benches mixed in with snack tables and cool coffee machines, and we always had music playing. The physical environment encouraged creativity and informal interactions, but at all times we demanded a high level of professionalism. We made it feel special, which made everyone at the lab feel like they were part of an important experiment, which they were."

- **No Boxes:** Andrew frequently used the expression, "We don't think out of the box, there are no boxes." Because many of the cybersecurity challenges confronted by the client were novel, there were few frameworks or standard algorithms to work from. Problem solving required truly original thinking, and that's where the trainees often excelled. Austin explained, "The young people we brought on didn't have cookie-cutter backgrounds and educations. They just thought very differently and often brought a highly pragmatic perspective." In the early days of the lab, the client brought forward a knotty cybersecurity problem related to a wireless technology that they had worked 6 months to solve, without success. One of the trainees, Damiete Roberts, turned over the problem with a completely different approach and presented a viable solution within 6 weeks. Andrew described the client reaction as "humbled and thrilled with the innovation." Here was a 23-year-old with no degree or years of prior cybersecurity experience, but with a bright mind unencumbered by preconceptions. Andrew is certain that this breakthrough was one contributor toward a large, single-source contract the client later awarded the lab.

- **Openness, Trust, Humor, and Encouragement:** Everyone in the lab felt they had room to succeed and room to fail. Successes were openly celebrated during the daily "stand-ups," and failures were held up in those same sessions as deep learning opportunities. For most of the trainees, it was their first professional environment and the first time they heard the word "brilliant" to describe their work and thought processes. Perhaps most compelling, each team member felt comfortable showing up to work as completely themselves, not an imposter of someone else they felt they had to be. It was not only the trainees who came from underrepresented and non-traditional backgrounds. Other diverse professionals who joined in full-time roles were attracted to the scrappy but welcoming environment where it felt like just about any achievement was possible. Out of the mission and camaraderie of the team came bonds of trust that created conditions for psychological safety, the concept pioneered by Amy C. Edmondson that we discussed in Chapter 1 and that is foundational to innovation.

- **Mentoring:** As we learned in Chapter 5, the impact of mentoring— especially on more junior employers—can be profound, but outstanding mentoring programs require strong support from leaders and constant fine-tuning. Even though Andrew was relatively new at Northrop Grumman, he quickly connected with senior leaders keen to see the lab successful and interested in the success of the trainees. Andrew matched each of the three trainees initially to executives responsible for large divisions or projects. While the exposure to senior leaders helped the trainees better understand Northrop Grumman's organizational structure and hierarchy, the interactions did not yield the deeper mentoring support Andrew envisioned. Realizing he needed to try another approach, he then matched the trainees to younger engineers who were one to two years out of college and situated their workspaces together. Borrowing a term from the U.S. Army (and coincidentally also used in Citi's apprenticeship program discussed in Chapter 4), mentors were called "battle buddies." Trainees interacted with their battle buddies daily on matters ranging from career tracks at the company to tackling detailed coding problems.

Moja Williams, one of the three initial trainees, shared a number of reflections during his months in the lab—what he now views as a pivotal time in his career. Moja enrolled in the NPower program after a family trauma disrupted his plans to complete a college degree. He had just finished his first year at Bowie State University in Maryland and was working as a lifeguard at the local YMCA when he learned about NPower through a friend. A big part of his motivation was driven by a desire to earn better wages quickly to support his family.

He described his first few days at Northrop Grumman as warm and inviting. He was initially fearful that his lack of coding skills would hold him back but was reassured by the step-by-step curriculum Austin curated. He describes his battle buddy, a recent Virginia Tech graduate, as an invaluable resource who was infinitely patient and helped him ramp up his coding skills, "not by taking over the machine, but by asking me a ton of questions."

When asked about the culture of the lab, Moja readily offered three observations:

"First, the Northrop Grumman team made all of us feel good about our work and about ourselves. There was a lot of humor and many jokes, but everyone also carried themselves very professionally.

Second, everyone could clearly explain the focus of the lab's mission and could break down how our jobs fit into the broader set of goals. We felt that what we did every day actually mattered.

Third, all of us who were trainees dearly wanted change. Yes, we loved technology, but we all saw the program as a way to do something very different with our lives. My other two peers both had two hour commutes each way but didn't complain because they knew how important this opportunity was."

Week 12: The Beginning of a New Employment Pathway

All three of the trainees were informed at the end of their 12th week that their contract assignments would be renewed at the end of the following month. Even better, their salary would increase by about 30 percent. The second contract would be for a minimum of 12–16 weeks.

This period of contract renewal gave Andrew and Austin time to finish their "secret mission"—to find a path to permanent employment for these first three trainees who did not have a 4-year degree, and for others who might come behind them. At Northrop Grumman, technology salary bands began at "T1," which is where recent college graduates with STEM degrees started, often as part of a rotational program. With the success of the lab and the trainees, Andrew and Austin proposed a "T0" band, or "Programming Support" title, for those coming from alternative pipeline programs who could be hired for full-time roles and put on a track to advance into T1 roles and above with corresponding salaries.

Astrid Portner, one of the more seasoned engineers hired for the lab, served as Miguel Velazco's battle buddy during his trainee tenure and became deeply passionate about hiring non-degreed talent. Austin described Astrid as one the brightest software engineers he's ever worked with, and she in turn used similar words to describe her mentee: "Miguel is brilliant. Even without a formal engineering background, he is superb at tackling tough problems. He just chews on a problem until he solves it, driven by initiative, hard work, and ingenuity."

A native of Baltimore County, Miguel thrived during his early years at school. He talks with pride about winning the President's Award for Educational Excellence in grade school. His life took a turn in middle school after his mom's divorce and financial hardship took him, in his words, "off the rails." By ninth grade he dropped out of school, and later passed the General Educational Development (GED) exam to earn his high school diploma. He worked several odd jobs to help his mom make ends meet, including as a cook, a "repo man" towing cars from people behind on their payments, and as a laborer erecting large party tents for events and weddings. Technology, though, was his passion. Like many people his age, he loved video games, but went further by teaching himself the coding required to build bots to play the games for him. From there he discovered great online content for learning cybersecurity skills. But he wasn't exactly sure how any of his self-taught skills would get him a job. His grandmother—who Miguel says always kept an eye on him—saw an online ad for NPower that led to his enrollment and subsequent job at Northrop Grumman.

Miguel distinguished himself during his contract period by combining strengths in spatial intelligence and technical knowledge with pragmatism. Austin tells the story of a ceiling-mounted projector breaking just before a high-stakes presentation to the client, "We had all these PhDs trying to fix the problem and we're getting more nervous by the moment. Miguel observes the whole situation then pulls up a chair, tinkers with the mechanics, twists it around, and boom—the projector is working."

How Success Ultimately Looked

Austin and Andrew were ultimately successful in getting a "T0" salary band authorized at Northrop Grumman. After a successful run serving out its mission, the lab came to a natural close and team members were reassigned to other divisions. Some colleagues across the company questioned whether the trainees who excelled in the protected lab environment that had a "culture within a culture" could succeed in other divisions, with other managers.

Yet, all three of the initial cohort of trainees thrived in full-time roles at Northrop Grumman, in addition to two others brought on board later and eight additional trainees (at the time of writing) in the pipeline for full-time positions. Those in full-time roles have received security clearances or are in the clearance process.

Moja Williams transitioned to a role as a cybersecurity analyst within the Mission Systems Department. His new boss values his experience at the lab and believes there were valuable lessons learned that could be applied across the company. Moja feels supported and hopes to continue to grow and develop into more senior roles at the company. And, thanks to the flexibility offered by his manager, Moja is one semester away from completing his bachelor's degree in computer science from Morgan State.

Miguel Velazco was selected to join the highly competitive "red team," which is charged with adopting the mindset of malicious actors who could breach the company's digital security systems. The ability to avoid confirmation bias or groupthink is tremendously important in the role of cyber protection, and Miguel's unusual background and approach is ideally suited to "thinking without boxes."

Andrew Parlock received the Greater Baltimore Committee's Bridging the Gap Achievement Award for his efforts at Northrop Grumman.

While he is now an executive with another technology company, he remains proud of the "high level of nerd and high level of ethics" he instilled in the lab.

Austin Cole has been promoted to a role within the Space Systems Division and has been tasked with building the next evolution of the Emerging Technology Innovation Lab, fueling a new channel of transferrable technologies and unconventional talent.

Ultimately, the grand experiment of the lab proved how one of the world's most complex companies forged a community and culture of deep curiosity, identified and nurtured talent from non-traditional backgrounds in ways that squarely met business requirements, and succeeded with an important client by suspending conventional biases.

Like the team at Northrop Grumman, the chief information officer at Tessco was keen to create opportunities for Baltimore residents within his fast-growing tech team. In the next case, we'll examine how this leader built in a career path from the first day of employment for new staff members without traditional technology or academic backgrounds.

Tessco: Reinstalling the First Rung of a Career Ladder

Tessco, founded in 1952 as an engineering sales and service company, delivers a range of technology products and solutions for companies that build, operate, test, and maintain wireless broadband systems. Their commercial clients are carrier and public network operators, tower owners, program managers, contractors, private system operators (including railroads, utilities, mining operators, and oil and gas operators), government agencies, manufacturers, and resellers. Product offerings include base station infrastructure, network systems, installation, and wireless maintenance products. Tessco is headquartered in Hunt Valley, Maryland, about 20 miles north of downtown Baltimore.

Jesse Hillman, Tessco's chief information officer, joined the company about a year after the arrival of a new CEO, Sandip Mukerjee. New leadership was on a mission to gain market share and improve profitability in a highly competitive industry sector. Jesse assessed that within his own organization he could contribute to overall corporate goals by containing costs and by hiring strong talent that could provide excellent technical service to Tessco's internal customers, who in turn supported external clients.

The Problem: Costs for the IT services that had been outsourced for several years were accelerating at a pace untenable for new budget targets. Further, by outsourcing IT functions, there were few pathways internally for developing mid-level and senior tech talent. Competing for experienced talent externally was becoming costly and time-consuming.

The IT Helpdesk Solution

How many times have you locked yourself out of your laptop? Or struggled with downloading apps on your smartphone? Perhaps you've lost connectivity with your company Internet? It's likely your very first call was to your company's tech support line.

People who staff the IT Helpdesk are the unsung heroes of digital transformation. Excellent Helpdesk professionals are good listeners, knowledgeable, fast, and patient. They are equally good at customer service as they are technically competent. In their roles they become experts in installing, maintaining, and repairing computer software and hardware and in troubleshooting any number of technical problems. They fill needed roles in virtually every industry sector, and often around the clock.

The job of Helpdesk Technician—also known as Field Technician, IT Administrator, Desktop Support, or Junior Engineer—is the first rung on the IT career ladder. Starting salaries are relatively modest, but the job provides a gateway to any number of roles in software engineering, cybersecurity, cloud computing, data science, and project management that often pay well over $100,000. It is the kind of "foot in the door" role where a motivated young person who is willing to work hard and eager to learn can forge a thriving tech career. In fact, many of the very senior chief information and chief technology officers we have worked with over the years began their careers as Helpdesk Technicians.

When these roles are outsourced to third-party vendors, a company effectively strips away the first rung of the IT career ladder for their internal staff. This is not to suggest that outsourcing is a poor practice. There are any number of good reasons outsourcing the Helpdesk function is often necessary, including access to trained talent, flexibility, cost considerations, and the need for 24-hour availability. But there are consequences that include reducing or eliminating the pipeline of in-house talent available for development into more advanced roles. Some third-party vendors have addressed this issue by providing customized training and permitting clients to convert contract talent to full-time employment, as we saw in Chapter 4 with the apprenticeship partnership between Citi and WWT.

Jesse saw an opportunity to "in-source" Helpdesk jobs at the company. He was concerned about the escalating costs associated with outsourced talent and wanted what he described as "a multi-prong approach to sourcing talent." He was also keen to provide career-path jobs to individuals who shared many of the same struggles that he experienced growing up.

As importantly, Jesse believed his organization would benefit from greater diversity along multiple dimensions. Jesse had been invited to join the Diversity Council at his prior employer. Leaders there knew that as a disabled veteran he would provide a unique and valuable perspective and Jesse believes he in turn learned from his peers. He attributes his personal values on diversity in part to his experiences as a naval intelligence officer where he witnessed professionals from virtually every corner of the U.S., and with varied socioeconomic, racial, and identity characteristics, come together to flawlessly execute complex missions.

A Career with Upward Mobility

Jesse conceived of a talent recruitment and training plan designed to attract new team members with a foundational level of technology knowledge, but not necessarily related job experience. He understood that one of the greatest barriers for those without a college degree, connections, or years of experience is just getting a foot in the door. Jesse believed that partnering with a Baltimore program that provided basic-level skills and industry-recognized tech certifications would be the right first step toward sourcing entry-level candidates. Jesse did considerable research on a range of programs and identified NPower as a good fit based on its technical skills and professional development curriculum. He confirmed his instinct after delivering a guest lecture where he got a chance to meet the local Baltimore staff and students.

From the outset, Jesse constructed a development plan for the NPower graduates he hired that would provide the additional skills and hands-on experience needed for advancement. Like some of the up-or-out style rotation programs for college graduates adopted by many large consulting companies, Jesse summarized a similar approach, "I don't want the people I hire through this program to be on the Helpdesk long-term. The goal is to assess in eighteen months whether (a) they can take on new, more advanced responsibilities at Tessco, or (b) if it's better that we help find them a role outside of the company where they can continue to grow. Either way, they will be well positioned to excel in tech careers longer-term."

With this approach, Jesse reinstalled the first rung of the tech career ladder for new entry-level talent at Tessco and created his own pipeline for mid-level technologists.

Building a Successful Talent Incubator

Since his objective was to cultivate an internal pipeline of talent who could grow with the company, Jesse was very intentional about crafting the first 18-month assignment to include technical competencies required to be successful in the Helpdesk position, as well as opportunities for new hires to gain confidence, learn about Tessco's policies and practices, and gain higher-level tech skills. Jesse gave this overview of his version of a work-based curriculum for new hires of the program:

- **Intensive Training:** For the first 45 days, new hires received 6 hours of training daily that included customer service protocols, Tessco's ticketing system, call prioritization methods, and specific upskilling in the platforms and infrastructure systems used by Tessco staff, who were clients of the Helpdesk. After a couple of weeks, new hires were taking calls and troubleshooting tech issues alongside a more seasoned staff member.

- **Weaving in Hands-On Experience:** Training persisted for another four and a half months, but with less intensity. Some portion of every day was dedicated to introducing a new security tool or software upgrade. During those months, trainees spent more time during the day answering Helpdesk calls.

- **Independent Problem Solving:** After the first six months, trainees were working independently to resolve increasingly complex tech problems. A manager was always nearby to provide support for especially challenging inquiries.

- **Self-Guided Learning:** Every trainee was given four hours per week to learn a brand-new skill that could lead to a more advanced assignment after the first 18-month period. Some studied data analytics, others pursued additional training in cybersecurity or cloud computing, and all were allowed to pursue topics of greatest personal interest.

- **Connecting Helpdesk Success to Business Success:** Throughout the program trainees learned about the Tessco business and the roles their internal customers performed across the organization. Many of the calls to the Helpdesk came from the Sales and Customer Success teams. Trainees were coached on how successful resolution of their clients' tech dilemmas directly impacted customer satisfaction and the bottom line.

- **Reviewing Performance:** Trainees were regularly reviewed and given straight feedback on strengths and weaknesses. There would be no surprises by the end of the 18-month assignment as to whether employment with Tessco would be continued.

Like many Helpdesk operations, Tessco uses a ticketing system that tracks the volume and nature of calls, and successful closure rates. Helpdesk staff were trained on how to log call outcomes that could be verified with internal clients. Performance was measured at the individual staff level against goals for number of calls successfully resolved per day. Jesse recalls a time when a new director reviewing the analytics suggested that two of the trainees be fired for not performing against goals. Jesse didn't hesitate with his response. He stood firm, reinforcing the purpose of the program, and emphasizing that he committed to each trainee an 18-month experience and did not expect perfection. If two individuals where underperforming, he believed they could be coached for improvement, which they were and remain successful at the company. And, the director who questioned the performance has since become one of the program's biggest supporters.

Jesse's view is that the individuals selected for the program—mostly young adults without college degrees and from some of Baltimore's most underserved communities, and all minorities underrepresented in tech—made a choice to invest in themselves and were unwavering in their ambition to succeed. Jesse observed what he described as "fortitude and determination" in each of the individuals selected for the program.

Partway through the program, one of Jesse's key team leaders unexpectedly left the company, which opened up a management role. With Tessco's overall growth and increased Helpdesk demand, Jesse split the job into two new roles and offered one of the positions to Rasheena McConneaughey, who joined as a trainee just seven months prior, well before the conclusion of the 18-month cycle. Rasheena had quickly distinguished herself with excellent customer service and surpassing her daily ticketing goals. She became a mentor and coach for the other trainees and emerged as a leader among the group.

The goal of developing a pipeline for more senior talent was getting met far sooner than ever expected. And all six of the individuals selected initially for the program remain with the company and—as of this publication—are preparing to take on more advanced jobs within the firm.

Rasheena, in her new management role, will take on the additional responsibilities of fine-tuning the training program and introducing a mentoring component for the individuals who get hired next. She thinks critically about the components of the program that worked well and those she wants to improve. In reflecting on her very first day at Tessco she remembers the sensation of seeing her cubicle with her nameplate, equipment set up and ready to go, and a backpack filled with Tessco goodies. "For many of us it was the first time we had jobs in an office environment, and here we had our own space with our names marked. We all instantly felt like we belonged and that the company was happy to have us aboard."

Today, Rasheena sees herself with a career at Tessco and is full of enthusiasm for, as she says, "easing the transition to the company for those coming behind me and guiding them to success."

For Jesse's part, he sees the program as helping the bottom line of the company and concretely advancing Tessco's commitment to diversity. He believes what he launched is a way for the company to "walk the talk" on diversity. He summarizes this talent development innovation as good for the company and for the trainees, all of whom are on career and income trajectories far steeper than in their pre-tech, pre-Tessco lives.

Now, let's examine how both the case studies of Northrop Grumman and Tessco align with our principles of innovation (Table 6.1) and then look at potential threats and how to prevent them (Table 6.2).

Table 6.1: Aligning Talent Sourcing Strategies with Principles of Innovation

PRINCIPLE NECESSARY FOR A CULTURE OF INNOVATION	PRINCIPLE IN ACTION
Courage	▪ Challenged a long-standing practice of out-sourcing IT Helpdesk support functions. ▪ Pioneered new model to transfer innovative technology developed in an unclassified environment to a classified category. ▪ Hired individuals with academic and work backgrounds different from most incumbent employees.
Risk-Taking	▪ Staffed a major client project with talent viewed as "unproven." ▪ Continued refining program to hire talent from non-traditional sources even after the first candidate didn't work out and initial efforts were criticized. ▪ Challenged long-ingrained systems of recruiting and hiring.
Trust	▪ Gave trainees the freedom to pursue independent learning on the job in technology topics of greatest interest. ▪ Provided room to both fail and succeed, emphasizing to trainees that learning comes from both outcomes. ▪ Established a cadence of frequent reviews for trainees, recognizing that honest and direct feedback builds trust and supports performance.
Collaboration	▪ Creatively worked with the training program's instructional staff to identify candidates who could excel in an engineering career path. ▪ Formed new partnership with HR teams to develop "work-arounds" to standard practice of offering work-based learning opportunities only to college interns.

Continues

Table 6.1 (*continued*)

PRINCIPLE NECESSARY FOR A CULTURE OF INNOVATION	PRINCIPLE IN ACTION
Leadership	▪ Relentlessly focused on culture and called out behaviors that contradicted an environment of inclusivity.
	▪ Understood that very few job candidates are "Ready on Day One" and that even the most experienced talent requires some form of onboarding and training to be productive.
	▪ Set a tone from the first day of work where trainees felt welcomed; "table was set" for their success from the beginning of employment.
	▪ Built successful programs that can grow over time in current division, and replicated in other parts of the organization.

Table 6.2: Potential Threats and Preventative Measures

INNOVATION THREAT	ACTIONS TO AVOID THREAT
Low Priority	In both the Northrop Grumman and Tessco cases, leaders actively prioritized sourcing candidates they could skill up for roles that were difficult and expensive to fill and cultivate these candidates to grow into mid-level positions. Leaving positions open and exceeding budget constraints were not options in either case. Both need and urgency fueled creative talent development approaches that succeeded.
Inertia	In the case of Northrop Grumman, Andrew and Austin's idea of "No Boxes" unleashed bright minds to challenge standard approaches to addressing the needs of an important customer and invent new solutions that were well-received.
	At Tessco, Jesse reintroduced the first rung of the career ladder, allowing entry-level talent to flourish in gateway roles. In doing so, he actively created a pathway for internal talent to grow into more senior jobs.

INNOVATION THREAT	ACTIONS TO AVOID THREAT
Arrogance	When the first candidate did not succeed in the new Northrop Grumman talent program, leaders did not see this entirely as a failure of the candidate. Rather, they questioned their own requirements for the position and chose to refine the skills and qualities they needed.
	Leaders at both companies approached the need for mentoring and well-designed onboarding curricula with humility and understanding. Importantly, these leaders were all at some level intrinsically motivated by their own personal early-career and life experiences and intentionally chose to invest personal energy into supporting the success of talent recruited from non-traditional pathways.

Conclusion

Leaders at Northrop Grumman and Tessco both chose to pursue unconventional solutions to their talent challenges and budgetary constraints. In the case of Northrop Grumman, Andrew Parlock and Austin Cole needed entry-level talent with foundational technology knowledge, the capacity and willingness to quickly learn software skills, and an abundance of unconstrained creative thinking—not necessarily college degrees. They needed people with a certain fearlessness. At Tessco, Jesse Hillman was challenged to find mid-level tech professionals and landed on the notion that he could successfully build his own talent incubator. In both cases, they found their solutions with young talent sourced from local Baltimore communities far too often overlooked as a pipeline for smart and deeply committed professionals.

Both companies invested in additional upskilling and mentoring for individuals sourced from a local tech training nonprofit to fully prepare them for their first assignments, but also to prepare them for continued development as the new hires demonstrated success. Both efforts set forth a goal to build career pathways that went far beyond entry-level tech support positions. And, while there was a need for upfront investment in time and resources to skill up the new talent, quality of performance and retention have supported continued investment with a generous return.

The experience of the leaders at Tessco and Northrop Grumman suggested that virtually no new hire—regardless of experience and education—is fully ready to be productive on "Day One." They both landed on creative solutions that challenged Fixed Attitudes and Fixed Practices to successfully identify and hire highly motivated individuals from diverse backgrounds who often traveled non-traditional educational pathways.

Summary

- Hiring based not on degrees, but on aptitude and skills—together with having the systems in place to build missing skills—can yield great candidates when requirements for success are clear and investments are made in supplemental training, onboarding, and mentoring.

- In a culture with a tradition of hiring only college graduates, it is especially important to create a sense of belonging and a welcoming environment for those hired for skills and competencies to help avoid imposter syndrome.

- Evaluate what the "first rung" of the tech career ladder is in your organization. Do you have accessible gateway jobs that can lead to development pathways into mid-level jobs where talent is especially challenging to find?

- A well-crafted onboarding curricula that sets aside focused time for both training and hands-on experience—together with a strong introduction to the core business—supports long-term success and retention of the new hire. Assumptions that an individual will sink or swim on their own will inevitably lead to failure and force a cycle that prevents new talent acquisition strategies from seeing the light of day.

CHAPTER 7

Innovations for DEI in Small Business

In 2021, the United States Small Business Administration (SBA) reported that there were 32.5 million firms that had fewer than 500 full-time employees, which represents nearly 99.9 percent of all U.S. businesses. Over 48 percent of the adult population is employed by a small business. Further, small businesses create over 66 percent of net new jobs and account for 44 percent of total U.S. economic activity. The idea that small business is a key driver of the U.S. economy is alive and well.

Launching a business—and even further, a successful one—is no small feat. Much of an entrepreneur's care and consideration must go into meticulous planning, market research, and competitive analysis, as well as effectively launching and growing the business. Doing a general search of "how to start a business" on a search engine or on popular retailer Amazon will yield no shortage of articles and guides on the subject.

What is perhaps harder to find, though, is guidance on how a new or small organization with limited resources can effectively create a meaningful DEI program, or even why they should care about DEI at the nascent stage of their business's journey. There is no doubt that there are many concerns that need an entrepreneur's attention in the short term, but embedding DEI early in a business's founding can be integral to its success over the long term.

In this chapter, we will focus on the unique challenges that small businesses face when starting DEI programs, and examine the DEI journey of Online Optimism in the aftermath of the killing of George Floyd. We'll take a close look at their journey to apply innovation principles to launch and strengthen its DEI efforts. The chapter also includes a story worth sharing on the owner of the Wesche Company, a small business based in Arkansas, and his personal journey as the catalyst for DEI in his company. Lastly, we'll hear advice from the owner of MojoTrek for low-cost, practical steps any entrepreneur can take to build DEI into their firm.

Key Concept: DEI programs and initiatives need not be limited to larger organizations. Small businesses can equally benefit from innovative approaches to building a more diverse and inclusive workforce as part of their overall competitive strategy.

The Challenge for Small Businesses Implementing DEI Programs

In a wide-ranging, two-part survey conducted by online business incorporation company Incfile, we see that there is a deep desire to do more for DEI, but that there are challenges holding back small business owners. The survey took a deep dive into the small business owners' viewpoints and their respective investments in DEI. The respondents were from various industries and 95 percent reported having less than 10 employees in their organization.[1]

The study revealed that 58.4 percent said that diversity plays an incredibly critical role in their business and their intention to make DEI efforts a priority. Fifty-two percent of the respondents also reported that they have integrated DEI initiatives in at least 30 to 50 percent of their overall business operations.

The study also underscored that many respondents believe that there is much work to be done. While 65 percent identified themselves as part of a racial, ethnic, or sexual minority group, almost 80 percent reported that they have not invested any funds in DEI programs or initiatives. Seventeen percent of respondents revealed that they would like to be

---

[1] Williams, W. (2022, July 18). More small businesses want to be diverse. So, what's stopping them? - Incfile DEI survey, from www.incfile.com/blog/small-businesses-diversity-areas-of-improvement

far more diverse than their present state. When asked why DEI wasn't more of a priority in their organization, the main reasons included:

- Their workforce isn't diverse enough now to have a DEI policy in place (40 percent)
- They are focusing on other priorities, such as marketing and sales (32 percent)
- They are investing in technology (29 percent)
- They don't understand how DEI impacts the business (18 percent)
- They don't know where to begin (17 percent)

This paints a picture of small business owners concerned about pressing business issues that are often perceived as disconnected from DEI practices, and who at the same time want to "do better" with DEI as a vaguely "good thing." We see a straight line connecting those urgent business concerns, such as cost and quality of labor identified in the Incfile study with steady DEI progress. There is as much of a business case argument for small businesses to adopt DEI practices as there is with their much larger counterparts.

More encouragingly, many entrepreneurs and business owners do draw this connection and see the value and merit of DEI for their organizations but are largely unaware of how or where to start, and how to operationalize DEI within their organization with their limited time and financial resources.

We've acknowledged from the start that implementing strong DEI programs and initiatives is difficult for businesses of *any* size, not just small businesses. From a cost perspective it can seem overwhelmingly prohibitive and restrictive to get *started*, let alone affording opportunities to apply innovation. As we mentioned in Chapter 1, the global spending on DEI is expected to reach $15.4 billion by 2026.

According to a 2021 Society of Human Resources Management (SHRM) report,[2] Fortune 1000 companies—the 1,000 largest American companies ranked by revenue—spend anywhere from $30,000 to $5 million per year for their DEI efforts, including training, recruiting, and associated programming, with the average spend of $1.5 million.

For small businesses, even at the smaller end of this range, this can be a hefty investment. The average annual revenue of small businesses in the

---

[2] Society of Human Resources Management. SHRM State of the Workplace Report 2021-2022. www.shrm.org/hr-today/trends-and-forecasting/research-and-surveys/pages/shrm-state-of-the-workplace-report-.aspx

United States, per the SBA, was $46,979 in 2021. As we've also acknowledged, the amount of money and time spent on DEI does not necessarily correlate to results, e.g., a more diverse talent pool, supplier diversity, etc.

Cost is not the only challenge facing small businesses that want to do meaningful DEI work. For many small businesses, remaining competitive against larger rivals, as well as attracting talent, expanding operations, and ultimately building profit, are just some of the many concerns that they face. And somehow, they must accomplish these goals often with a limited number of staff and resources.

Marina Perla, chief executive officer and founder of technology staffing firm Mojo Trek—which is also a certified Women Owned Small Business (WOSB) through the Women's Business Enterprise Council—understands firsthand the challenge of prioritizing with limited resources. "I don't know if that's unique to small businesses, but the pressure to grow, acquire clients, hire the best talent without having access to financing or impressive cash flow is definitely stronger," she says. "At Mojo Trek, we have an ever-long list of projects and competing priorities. We know we cannot do it all and have to prioritize every year."

Most U.S.-based small businesses have less than 10 full-time, salaried employees. An October 2022 survey by small business advocacy organization National Federation of Independent Business (NFIB) revealed that the most important problems for small businesses included inflation, quality of labor, taxes, and labor costs, but not efforts on DEI.[3]

DEI – Not Impossible for Small Businesses

Yet embedding and prioritizing DEI from the start can stave off many of the Fixed Attitudes and Fixed Practices that a company can develop over time, as we discussed in Chapter 3.

"I believe it's easier to build it right from the beginning [than] to try and change bad habits and biases once they become ingrained in the way that a company does business. Think about it as a grassroots approach. When the team is new and individuals grow as the business grows, they are more adaptable and open to changes. Even if DEIB [B for 'belonging'] wasn't part of their agenda prior to joining the business, they learn from the new team and leadership quickly. . . and they start

---

[3]NFIB Research Foundation. (2022, November 8). Small business economic trends, from www.nfib.com/surveys/small-business-economic-trends

integrating the same approach in their work. They truly buy into that and become advocates for diversity and inclusion," Perla notes as she reflects on how she has prioritized diversity at Mojo Trek.

In our discussions with small business leaders, while the challenges are plentiful, they are not insurmountable or impossible to work through. There are opportunities for small businesses to create impactful and meaningful programs, utilizing existing resources, ingenuity, and help from others. Let's see this in action in the case of Online Optimism.

Setbacks and Progress: Online Optimism's DEI Journey

Launched in 2012, Online Optimism is a creative digital marketing and design agency based in New Orleans, Louisiana. Flynn Zaiger, chief executive officer, founded the company soon after graduating from Tulane University. What started as a solo endeavor operating out of his home soon became a 20-person company, sprouting additional offices in Atlanta, Georgia, and Washington, D.C. Of its current 20-person staff, approximately half identify as Black, Indigenous, and people of color (BIPOC).

Zaiger and his leadership team have made it a priority to create a positive and inclusive culture for all staff, who refer to themselves as "optimists." The company's internal Culture Committee, made up of Online Optimism staff of various levels, was established to constantly find and explore inventive ways to bolster employee engagement, especially in light of their expansion beyond their New Orleans office. The company website transparently details their mission, vision, and values:

- Build on trust
- Be exceptionally helpful
- Keep it human
- Always optimize
- Support our communities[4]

The intervening years brought many additional clients, as well as accolades, including a *Best Places to Work* honor from *New Orleans CityBusiness*

---

[4]Mission, vision, values. Online Optimism. (2022, September 28), from www.onlineoptimism.com/culture/mission-vision-values

magazine in 2015. In 2021, the company received the Best-in-Class Gold Health @Work Award from ComPsych, a major provider of employee assistance programs, for its commitment to the health and well-being of its employees.

Turning Point: Developing a Deeper Appreciation for DEI

When speaking with Zaiger specifically on his and his company's path to DEI, he was candid on the driving force: "mostly mistakes that we've made," he says. "I think that if I started the company now, I would have more of a sense to think about this early on. But as a 22-year-old entrepreneur. . . it just wasn't at the forefront of my mind."

For many entrepreneurs, keeping adequate cash flow, securing capital, developing a marketing strategy, and landing new clients are just a handful of the many challenges that they face daily, as we established earlier in the chapter. That alone would make most people empathetic and sympathetic to the plight of entrepreneurs, and perhaps willing to give a "pass" for not prioritizing DEI. Zaiger saw this differently, saying "When people start a business, a lot of times, they don't really think about these things, and that's not really an excuse."

The Problem: Creating a meaningful and impactful DEI program in the face of constrained resources, limited existing knowledge, and unprecedented social unrest.

New Orleans has historically been a city that has struggled with crime, poverty, and significant environmental events and risks. A disproportionate number of residents of color have felt the effects of these challenges. Zaiger says as a business in the heart of the city, it's almost impossible not to acknowledge the racial and economic disparities.

According to the 2020 U.S. Census, of the more than 370,000 people that reside in the city, 59 percent of the population identified as Black or African American compared with 33.4 percent that identified as white. Looking through the lens of employment alone, only 53 percent of Black/ African Americans were employed, while 78.2 percent of those who identified as white were employed.

In this context, DEI work is particularly critical. Zaiger recounted a client project that shed light on opportunities for growth in 2016 that led to his own deeper appreciation for internal diversity. The company was hired to take on several marketing activities for a local, federally qualified health center (i.e., a healthcare organization that provides medical services to help an underserved area or population). The client took issue with the language the company chose to use in the creative work it presented, which the client pointed out was not in line with marketing and advertising that would resonate with the communities who were the audience for the campaign, and could potentially be construed as insulting.

The client insisted that leaders from Online Optimism attend anti-racism training to better understand and appreciate the challenges of the communities served by the health center. This training was not optional and was a requirement from the client in order to keep working together.

"It's not a real winning story, to be frank, to be told we had to go [to anti-racism training], but honestly that was how the conversation started," Zaiger says. And certainly, not something that's helpful from a business's financial bottom line.

Per the insistence of their client, Zaiger and a member of his team attended a 20-hour anti-racism training course provided by The People's Institute for Survival and Beyond (PISAB). PISAB was founded in 1980 by Dr. James Norman Dunn and Ronald Gunn, with the goal of helping to cultivate anti-racist communities. In addition to training, the organization also offers consulting services and workshops to businesses and communities alike.

Zaiger could not comment on the exact details of the training (part of the training requirements ask that attendees not publicly discuss details of the workshop or their conversations). The training's goal is to help attendees understand the history of race in America and how it affects individuals, personally and professionally. Zaiger revealed that the training "became the forefront of most of my decision making, which I think goes to show the impact of that sort of workshop."

While the workshop helped him and his colleague see how little knowledge, understanding, and training the team had on these matters, it helped to initiate some of the tougher conversations on race and racism that needed to happen internally, and ultimately, to minimize the possibility of mistakes of language, cultural insensitivity, and misrepresentation in future campaigns.

Blackout Tuesday: Paving the Road for Meaningful DEI Innovation

Zaiger noted that while he found the anti-racism workshop from PISAB incredibly enlightening, the company did not undertake company-wide DEI training immediately afterward. "I don't think I led as well as

I should have," he says. "I was a lot more knowledgeable of it and kept an eye out for [DEI], but we really didn't implement any training until after 2020."

What would prove to be a catalyst for training was Blackout Tuesday. To protest racism and police brutality, and to encourage the larger global population to reflect on the aftermath of George Floyd and the longstanding institutional racism that Black communities have endured, some businesses opted to go on a "blackout" the summer of 2020. During blackouts, businesses would refrain from their usual commercial activities to bring awareness of these issues to others, in the hopes that they would spark reflection and frank dialogue.

There has been much criticism of this approach to social activism because it is often viewed as an example of performative allyship. An example: Many organizations, on their social media accounts, posted a black square image and #blackouttuesday hashtag, but often with the poster offering no further discussion or dialogue about racism and what, if anything, they were committing to do in their own lives or businesses to remedy the situation. Additionally, many critics felt that by also using the #blacklivesmatter hashtag in these posts, the impact of the actual Black Lives Matter movement and the need to share critical information was diminished.

Zaiger recalled that he had been on vacation at the height of Blackout Tuesday and that a staffer reached out because they believed that the company should *do something*. This led to a reflection on the fact that the company still did not have formal DEI programs or policies, nor had it conducted any trainings or substantive internal conversations on DEI. Zaiger did not want to follow in the footsteps of other companies; if they were going to react, he wanted their work to be serious, intentional, and meaningful, and not just a "one-and-done" social media post.

It was from there that the company decided to take on several DEI initiatives that would affect the organization and its immediate community. On June 10, 2020, the company posted a blog post titled "Black Lives Matter," which was also distributed by email. The blog post not only called out the performative nature of Blackout Tuesday, but the actions that the company intended to take, including:

- Making financial donations to local organizations dedicated to helping underserved populations
- Investigating partnerships that will leverage the company's digital marketing and advertising expertise to amplify the work of Black-owned businesses

- Enrolling employees in anti-racism training
- Improving the diversity of hiring by partnering with organizations like Historically Black Colleges and Universities (HBCUs)

The company has often updated the blog post page from 2020 into 2022 with its progress and new initiatives, such as a book club and using more inclusive stock imagery created by BIPOC talent (like the image site Macro).

Today's Outcomes, Leading to Future Impact

Zaiger commented that it has been two years since an official annual update to this post was made and that part of this stems from the fact that he doesn't feel like they are doing 100 percent right, even though he knows "not that anyone is doing it perfectly." An example: One of the goals highlighted in the Black Lives Matter post is to have staff that "visually reflects" the city they hire from. At the time of the post's writing, only 7 percent of the Online Optimism employees identified as Black. Here, Online Optimism intends to do deeper outreach to HBCUs and other organizations that can help identify top diverse talent for its pipeline.

Still, its progress in the DEI space is commendable and it has made an impact both internally and externally. All employees now go through formalized DEI training, including paid interns. The company has employee affinity groups, including those for the BIPOC and LGBTQ+ communities. Its *On The Same Page* book club selections have included critically acclaimed readings on race such as *So You Want to Talk About Race* by Ijeoma Oluo, *The Fire Next Time* by James Baldwin, and *How to Be An Antiracist*, by Ibram X. Kendi, all of which have inspired reflective and authentic conversations about equity in Online Optimism's own workplace, and how it can best serve its internal and external communities.

Employees have also attended trainings on how to better reflect diversity in marketing and advertising, and to create inclusive campaigns. Some of its client success stories include work for minority-owned businesses, including Space to Reach, a recruiting platform for Black Women in STEM professions, and K. Allen Consulting, an education and management consulting firm.

In 2020, Online Optimism, along with other local area creative agencies, issued their first annual V.I.B.E. (Venture to Ignite Black Entrepreneurship) Creative Marketing Grant, in an effort to help Black-owned businesses in their marketing efforts. The grant includes a $5,000 paid digital advertising budget and a robust print and media campaign for the recipient.

In addition, for donations through its *Donate, Elevate* program, where the company provides a 100 percent match to employees' non-profit donations, the company also participates in team volunteer opportunities and amplifies the work of area nonprofits on its website. An example: In early 2021, the company participated in a day of service for the Sankofa Food Pantry. The organization offers several meal assistance and economic development programs for the Lower Ninth Ward, an area that experienced tremendous devastation in the wake of Hurricane Katrina in 2005.

MAKING IT PERSONAL: THE WESCHE COMPANY'S PATH TO DEI

A family-owned business based out of Springdale, the Wesche Company of Arkansas is a distributor of commercial doors, frames, hardware, and building specialties, serving the commercial construction industry in northwest Arkansas. Founded in 1996 by his parents, Jacob McConnell became president in 2020 and now manages the company's day-to-day operations. There are currently 30 employees at the company, where approximately three employees identify as diverse talent.

Like many leaders, McConnell believes in the business value of DEI. He made note that while northwest Arkansas does not boast a diverse population (in 2020, nearly 80 percent of the population identified as white), he also noted that there is a large and growing Hispanic/Latino community, and many of them work in the construction industry. "For us, the more that the office can reflect our customer base, I think that's good for us. When someone comes in and only speaks Spanish, us being able to have salespeople who speak Spanish and identify with our customers, will make us a place that I think customers will enjoy coming back to more."

It was a unique opportunity with the nonprofit Teach for America that really drove the importance of DEI on a personal level for McConnell. After graduating from the University of Arkansas, McConnell had the opportunity to teach in rural Louisiana. "I kind of had my eyes open as to just the huge inequitable situations in our country, based off of where you were born and what your race is." The experience also introduced him to his wife, who is a Black woman. McConnell recognizes that his is a unique, "not universally translatable" experience and that even if others don't agree with DEI from a moral or ethical perspective, for him, it's deeply personal. "My children are bi-racial, my in-laws are Black Americans and I recognize that my experience growing up is not that of a lot of other people's, and a large part of my religion tells me that I'm supposed to work on behalf of people who have less power and try to make things more equitable."

While the company did not have any formal policies, initiatives, or metrics surrounding DEI, the values were there from the very start of the business. McConnell stated that he had had conversations with his parents, who were in alignment with his moral drive to build an inclusive culture. "When hiring employees, we'd talk about it, or, wanting to find someone who could speak Spanish. It's not like the concept of wanting to do this was brand new since when I started."

McConnell would have the opportunity to create and put a formal DEI program in action starting in 2020. The Northwest Arkansas Council, which aims to help accelerate workforce and economic development in the region, launched the NWA Pledge. The pledge calls for organizations who took it to find ways to address systemic racism and forge a more equitable and diverse community. NWA educational and awareness activities began with textbook studies on race and instructional webinars.

The Council then developed a program partnership with the University of Arkansas IDEALS Institute. For companies selected for its partnership program, IDEALS (Inclusion, Diversity, Equity, Access, Leadership Development, Strategic Supports) hosts a 12-week cohort to help them develop a formal DEI plan for their businesses and build up their core DEI competencies. McConnell applied and the Wesche Company was accepted.

In addition to the development of a formal DEI plan, the cohort engaged in regularly scheduled meetups to discuss DEI topics such as microaggressions, the definition of equity, and how to conduct intercultural assessments.

The DEI plan produced for Wesche was comprehensive, including the development of a DEI statement, and the establishment of short-, mid-, and long-term goals. Short-term goals for the company included the creation of an internal DEI committee who will develop and implement DEI initiatives in all facets of the organization, aligning charitable donations to organizations where DEI is the focus of their work, and working with three local high schools in an effort to diversify Wesche's talent pipeline. More longer-term goals included identifying non-white, non-Christian-centric holidays for celebration and creating individualized training plans to ensure equitable opportunities for career opportunities and progression. The plan was discussed with the IDEALS staff and was presented to the cohorts from other companies for questions and feedback.

Reflections on Turning the Plan Into Action

The disruptions of the pandemic and uncertain economic conditions posed numerous challenges to implementing the plan. McConnell acknowledged that the last two years were difficult to navigate and that DEI was not always his central focus. "We deal with lead time issues, we're dealing with staff shortages. . . so there have been so many days where it's like operating in a mode to get through this week and put out these fires, that long-term DEI is a less hot fire in terms of what's pulling our attention on a day-to-day basis. While I'd love to say that I've carved out the time to think about how we're progressing towards these long-term goals, sometimes it's like, oh my gosh, my customer needs this product in three weeks and we can't get it for two months!"

The other major challenge that McConnell sees is building the momentum behind DEI when the organization itself is not diverse. McConnell recounted the development of the company's DEI committee, where he reached out to staff asking if there was anyone interested in participating. While there were volunteers, all of them were white. He felt uneasy asking the few people of color in the company to participate; he did not want to burden them or force them to do something that they didn't want do.

Another example of this dynamic came to light when focusing on the holidays not traditionally celebrated in the office. The company decided one year to celebrate Día De Los Muertos ("Day of the Dead"). The Mexican holiday, typically celebrated annually October 31–November 2, celebrates the spirits and souls of relatives who have passed away. In addition to spending money with Mexican-owned businesses, they set up tamales and pan de Muertos ("bread of the dead") for employees.

But McConnell felt uneasy. "I don't want us to be culturally appropriating, since at the end of the day this is all done by white people." He considered acknowledging Eid al-Fitr, which marks the end of the Muslim holiday Ramadan, but McConnell's wife (wisely) counseled that such a celebration would make no sense for the company to do. "Our intent is to celebrate other cultures and make it feel more inclusive. But at the same time, it would be us randomly thinking of what to do and could not be done well. I think getting the ball rolling in an authentic way in DEI is a challenge."

Still, despite his company's challenges, McConnell encourages small business owners to turn their focus to DEI early in their journeys.

McConnell also encourages leaders to reach out to others to have candid conversations on the topic. "I know that a lot of small businesses may not have boards, and they may not have wives like mine either, but find someone to. . . go with an idea and feel comfortable with being open and honest to give that idea and then say, 'Does this seem like a good idea?'"

Even with limitations, McConnell had the vision and courage to seek others' help in developing a meaningful DEI program his company, and trying different approaches to find initiatives that were important to the company's staff.

From Concept to Realization: Creating DEI for Small Businesses

For entrepreneurs and small businesses, starting a DEI program can be difficult, given limited funding, staffing, and perhaps, a lack of a local diverse pool to find talent. Yet, as we've seen in the previous cases, it is not impossible. Since Perla from Mojo Trek began her DEI program from scratch, she has documented their progress at every step and offers this counsel for small business leaders:

- **Commitment from the top:** Perla, on behalf of Mojo Trek, signed the CEO Action for Diversity and Inclusion Pledge. The initiative, launched in 2017, has over 2,000 CEOs committed to driving change in their organizations, and to cultivate more diverse, equitable, and inclusive environments.[5] Mojo Trek signed the pledge of August 2022. "Their purpose resonated the most with me," Perla explained. "Viewing DEI as a societal issue and not a competitive one and the idea that it takes all of us to make a positive change. I like the accountability factor—by making a public statement and adding our name and logo to the signatory list, we naturally create some positive pressure to keep DEI on top of our agenda and continue to foster this mindset within our organization."

- **Utilize free resources:** Perla also highlighted a key benefit from signing the pledge—access to free training and resources for her and her team to utilize. "Since we have a small team, we love, love, love the free trainings they provide on unconscious bias, sourcing for diversity, microaggressions, etc."

- **Staff support:** Within the company, a part-time team member has been tasked with ensuring that activities within the DEI program are executed and measured against stated goals. As Perla explained, they propose the plan for the next quarter/year, advise on what they think is possible to cover within that time period, and solicit recommendations from leadership to finalize their plan.

- **Assessing where you are and where you'd like to be:** Understanding which facets of diversity you intend to measure and cultivate is also important. "Diversity can shine through in so many different ways and faces," Perla says. "We can talk about age, gender, veteran [status] education, [and] representatives of the LGBT community." The company decided to focus on diversity by race and gender, which for the technology field remains, as Perla noted, a pain point. Currently, one of its goals is to raise the gender representation among its technical consultants from 37 percent to 45 percent.

- **Adopt inclusive recruiting and retention practices:** Adjusting the language or terms used in job descriptions can make a difference

---

[5]PwC. (2017). Purpose. CEO Action for Diversity & Inclusion, from www.ceoaction.com/purpose

in diversifying the candidate pool. Perla calls out dropping "nice to have" or "optional" sections, or degrees that are not required, from job descriptions, as they may disproportionately affect candidates who are Black or women. Adding phrases such as "flexibility" can also help increase engagement with diverse audiences, particularly those who, due to their personal obligations, need more autonomy on when and how they complete their work.

■ **Consider flexible work options:** Having flexibility on when and where workers can perform their jobs can help increase the pool of diverse candidates to hire, especially in areas where the local population itself is fairly homogeneous. Perla offered, "a job that offers a shortened work week or work day would be appealing to a wide variety of people, such as stay-at-home parents, international talent in distant time zones, caregivers, people in or nearing retirement, etc." For remote work, Perla understands that there are unique challenges, but also cites that it has helped her clients attract talent from well beyond their headquarters and was a net positive on improving their cultures. We'll speak more on remote work specifically in Chapter 8.

Table 7.1 reviews the approaches both our case study companies from this chapter took to integrate DEI into their small business. Table 7.2 reviews the potential threats and actions that may prevent them.

Table 7.1: Aligning Small Business DEI Strategies with Innovation Principles

PRINCIPLE NECESSARY FOR A CULTURE OF INNOVATION	PRINCIPLE IN ACTION
Courage	■ Online Optimism was courageous in addressing where its gaps in DEI understanding were and where it could improve. It pursued programs and initiatives that made sense for the company, and did not subscribe to performative allyship, even if it was perhaps the easier path.
Risk-Taking	■ Because DEI is not commonly an immediate area for development for small businesses, Online Optimism was in unchartered territory and did not have many peers to model its DEI efforts after.

PRINCIPLE NECESSARY FOR A CULTURE OF INNOVATION	PRINCIPLE IN ACTION
Trust	■ The leaders at Online Optimism placed a great deal of trust in those outside the organization to provide guidance in the development and refinement of its DEI practices.
Collaboration	■ Online Optimism not only worked with cross-functional leadership within the organization, but worked with local businesses and organizations to identify additional areas for improvement and community impact.
Leadership	■ The entire leadership team at Online Optimism is committed to DEI efforts in the organization. No one leader or one part of the organization takes on DEI responsibilities alone.

Table 7.2: Innovation Threats and Preventative Actions

INNOVATION THREATS	ACTIONS TO AVOID THREAT
Arrogance	■ Adopting a mentality of learning and humility helped the leaders at both organizations recognize that DEI is a journey full of continuous improvement.
Low Inertia	■ Since 2020, Online Optimism has recognized the importance of a sustained effort in DEI, with the setting of goals and holding leadership accountable for goals being met. By publicly announcing and documenting its DEI commitments and actions, it serves as a constant reminder that DEI requires long-term commitment; practices and initiatives change over time.

Conclusion

Online Optimism's journey in its DEI efforts is innovative in several respects and can serve as a model for other small businesses. In addition to applying the innovation principles, the company utilized several low-cost modalities to put its DEI learnings into practice and to apply the understanding provided by formal training into improved creative products and campaigns for its clients. Not only has the company succeeded in building DEI into the DNA of the company, it has integrated its brand of inclusion into a better business strategy that responds to a wider range of clients.

Participating in external and community volunteer opportunities not only allows for employees to make a tangible impact on the people that they serve, but it provides opportunities for genuine dialogue. Book clubs, in a similar vein to the listening sessions offered by World Wide Technology (more on that in Chapter 9), can give employees a deeper understanding of the struggles and plights of marginalized groups, and thus inform actions that leaders and employees can take in being more intentional in their DEI efforts. These approaches require only a commitment of time versus a commitment of funds.

Partnering with external organizations to better understand what DEI is and what it means to their organizations can help small business leaders create more meaningful programs. In the case of both Online Optimism and the Wesche Company, both leaders took the time to reach out to subject matter experts to better understand the underpinnings of racism, bias, and discrimination in the U.S. workplace, which helps to better inform their approaches to DEI.

Summary

- Small businesses often face more challenges than larger businesses when it comes to implementing DEI. Among them are costs, investing in other key areas of the business, not being diverse themselves, or simply not knowing where to start.

- Small businesses do recognize the importance of DEI, from a growth, profitability, and talent acquisition and retention perspective.

- Like many other organizations, Online Optimism realized it had gaps in its knowledge and understanding in DEI and created a more robust DEI program in the aftermath of George Floyd.

- Online Optimism's program employs many traditional strategies, like formal training, and also other approaches, such as volunteerism, book clubs, donations, and community outreach.

- Online Optimism created a bespoke, meaningful DEI program that made sense for its organization, and has avoided performative or hollow approaches to its DEI work.

CHAPTER 8

Rethinking Retention Through the Lens of DEI

While recruiting for diverse talent can be challenging, keeping and growing workers to their fullest potential can be even more daunting. Employers can be particularly challenged for Gen Z workers, who are more likely to place a premium on workplace culture than other factors when considering an employment opportunity.

As discussed in Chapter 1, many employees reevaluated whether they felt their professional and personal needs were being met by their current roles and employers during the COVID-19 pandemic. The answer, in many cases, was no, leading to the Great Resignation, where over four million employees voluntarily resigned from their jobs,[1] or engaged in "quiet quitting," where employees do the minimum amount of work required for their jobs and work solely during agreed-upon work hours. Here, we'll explore the factors that contribute to job turnover in general and how these are more acutely felt by underrepresented groups.

---

[1]Tappe, A. (2022, March 29). The great resignation continues: 4.4 million Americans quit their jobs last month, from www.cnn.com/2022/03/29/economy/us-job-openings-quits-february/index.html

We'll examine the areas of disconnect between what employees and employers believe make a good and inclusive workplace. We'll also examine innovative approaches to retaining diverse talent—including career advancement for experienced professionals, remote work, and employee benefit programs.

Key Concept: A variety of retention strategies must be considered and implemented to address the three primary reasons people most often cite for leaving their jobs. In order to strengthen an organization's equity efforts, retention strategies must consider differential impacts on various groups within the company.

Understanding Employee Retention and Turnover

No matter the industry or the size of the organization, employee retention—the ability to keep employees engaged—is critical for a business's long-term success. From a financial perspective, it's more than just the salary and benefits costs. There are also costs associated with recruiting, hiring, onboarding, and training a new employee. The Society for Human Resource Management ("SHRM") estimates that the national average cost to hire a new, full-time, non-executive employee in 2022, independent of salary and benefits, was $4,425, while the cost to hire a new executive was $14,936.[2]

This does not include costs to integrate new employees in their workplaces (e.g., the purchase of office furniture, supplies, and hardware/software needed to do the job), or the amount of time it takes an organization to complete the tasks associated with hiring activities. Nor does it include the costs of only partial productivity during the first several months any new employee requires to get up to speed in a new role.

With tangible and intangible costs in mind, it's just smart business for leaders and hiring managers to keep their employee turnover rates low. Turnover refers to the number of workers that leave an organization during a given time frame, be it in a month or within a year, and are replaced.

What constitutes an "ideal" turnover rate varies greatly by industry and by job function. For example, historically jobs in the retail industry have had a high turnover rate partially due to the seasonality of certain positions, but primarily because of the perceived lack of benefits and

---

[2]Upwork. (2022, September 29). The cost of hiring an employee: Explanation and formula, from www.upwork.com/resources/cost-of-hiring-employees

advancement opportunities. By contrast, jobs within state and federal governments have a below average turnover rate, as government employees are less likely to quit their positions due to the relative job stability and benefits that government jobs afford.[3]

Understanding the data behind employee retention and turnover (including the specific reasons employees cite when choosing to leave a role or organization) gives leaders indications of how effective their employee management and hiring practices are. But it also is a strong indicator of an organization's culture. It provides leaders with a meaningful barometer for just how happy, or dissatisfied, workers are with their roles, their teams, or their organization.

Contributing Factors for Recent Turnover Trends

While people voluntarily leaving their jobs is certainly not a new phenomenon, the rate at which resignations occurred during the Great Resignation was too extensive to ignore, especially in the wake of a global pandemic. Workforce participation since the pandemic has rebounded, but the experience did prompt employers to better understand why good people were leaving—even if they have been, historically, largely considered an employer of choice by candidates.

In SHRM's 2022 *Better Workplaces on a Budget: Recommendations for Retaining Employees Without Additional Spending* survey, over 1,500 U.S.-based HR professionals were asked the top reasons employees left their organization. The top three reasons were:

1. Inadequate total compensation—pay, bonuses, and profit sharing (74 percent of respondents)
2. Lack of career development and advancement (61 percent of respondents)
3. Lack of workplace flexibility (43 percent of respondents)

Companies that are innovating for diversity are addressing these three most frequently cited reasons for turnover with creative solutions and a focus on how policies may disproportionately benefit or disadvantage any particular group. Before examining these imbalanced impacts, we will look at each of these "pain points" in general.

---

[3]Lewis, G. (2022, August 11). Industries with the highest (and lowest) turnover rates. LinkedIn. www.linkedin.com/business/talent/blog/talent-strategy/industries-with-the-highest-turnover-rates

Compensation

On the surface, compensation may appear to just be "I need to be paid more" conversation. But underlying demographic and economic trends shape a more complex picture of employee concerns over financial health. According to a recent report by the workplace research firm Syndio, each generation is confronting a different set of financial pressures. Today, the majority of Baby Boomers (those born between 1946 and 1964) are still working, and the oldest remain in the workforce at higher rates than previous generations, largely driven by financial insecurity. Nearly two-thirds of Boomers are concerned about having sufficient savings to retire and over 20 percent are delaying retirement because the pandemic made them feel less financially secure.[4]

Gen X (born between 1965 and 1980) households have the highest income, but also the most offspring and the largest debt of any generation. They are concerned about rising health care costs as they care for both children and aging parents. Nearly half of Gen X'rs are supporting a parent or child over 18 still living at home.

With the majority of Millennials (born between 1981 and 1996) entering the job market during or just after the years of the Great Recession, they hold less wealth than those from previous generations at the same age and earn roughly 20 percent less than Boomers at the same stage of life, despite having more education. Student loan debt is burdensome, preventing Millennials from buying a home. In fact, almost half of all Millennials live paycheck to paycheck and 67 percent work a side hustle just to make ends meet.

Gen Z (born between 1997 and 2012), the most racially diverse and most educated generation, is more aware of issues like systemic racism and intersectionality. Those who are already in the workforce seek inclusive cultures but are also sensitive to financial pressures. This generation was hardest hit professionally by the pandemic: nearly half are worried about covering their daily expenses and 30 percent report feeling financially insecure.

Finally, at the time of this writing, the U.S. inflation rate was 7.7 percent[5] while the average salary increase in 2022 was 3.2 percent.[6] Workers of

---

[4]Martin, C. (2022, August 4). Workplace equity by generation: Millennial pay gap & more from https://synd.io/blog/workplace-equity-by-generation-stats-millennial-pay-gap/#:~:text=Millennials%20and%20Gen%20Z%20have,reject%20a%20low%20salary%20offer)

[5]Al-Shibeeb, D. (2022, November 10). Current inflation rate. Moneywise, from https://moneywise.com/news/economy/inflation-rate

[6]Miller, S. (2022, April 12). Revised 2022 salary increase budgets head toward 4 percent. SHRM, from www.shrm.org/resourcesandtools/hr-topics/compensation/pages/revised-2022-salary-increase-budgets.aspx

all generations experienced financial backsliding with less opportunity to build wealth in the period following the pandemic.

Limited Career Development or Advancement

Workers need to feel they have tangible advancement opportunities in their organizations; the lack of skilling or re-skilling opportunities was a primary factor in their departure. In the SHRM study, 21 percent of the respondents cited that this was their top reason for leaving an organization.

This finding reveals an alignment between the needs of any company for strong employee retention and performance, and the desire among employees to grow within a company. In a 2022 study, Deloitte found that engagement and retention rates of employees were 30–50 percent higher in organizations with a strong learning culture focused on supporting staff development. Importantly, they are also 92 percent more likely to develop novel products and processes, 52 percent more productive, 56 percent more likely to be the first to market with their products and services, and 17 percent more profitable than their peers.[7]

So with all the data suggesting that investment in training and development supports both retention and performance, why do nearly 60 percent of U.S. employees report lack of access to skills growth? Why doesn't every company invest in their employees? The short answer: too many leaders believe, "why invest in development when my employees are going to leave anyway?"

Companies that take a firm stand and prioritize employee development will be the winners in the 21st century.

Inflexible Workplaces

Options for flexible work schedules, where employees can change the hours in which they perform their work, as well as the possibility for remote work—allowing employees to perform tasks outside the traditional office—have emerged as primary considerations when workers contemplate a new opportunity. To effectively compete for talent, smart companies are adopting hybrid work policies that respond to the needs of both employees and business goals, investing in virtual collaboration and communication technologies, and training team members to interact across multiple platforms.

---

[7] Deloitte Insights. (2015, January 27). Becoming irresistible: A new model for employee engagement, from www2.deloitte.com/us/en/insights/deloitte-review/issue-16/employee-engagement-strategies.html

In its 2022 American Opportunity Survey, McKinsey reports that 58 percent of American workers now have the opportunity to work at home at least one day a week. Of those given the option to work flexibly, 87 percent take it. Interestingly, respondents to the survey work in a range of job categories, and in every region and every sector of the economy, including "blue collar" jobs that traditionally require on-site labor, as well as "knowledge" professions.[8]

Retention, Turnover, and DEI

Now let's consider the impact when compensation equity remains elusive, opportunities for advancement favor majority populations, and stigmas surround flexible work schedules endure. In particular, we'll look at the issue of career advancement in considering retention and turnover among Black professionals.

When we look at retention and turnover through the lens of DEI, we find that leaders must pay even closer attention to providing the right levels of support for underrepresented groups to advance.

McKinsey's 2021 *Race in the workplace: The Black experience in the US private sector* survey highlighted where the challenges persist for Black employees. Of note, while companies were effective in bringing Black employees to early career positions (jobs that require less than three years of professional experience after high school or college), they were far less effective in promoting them to managerial or senior levels. The participating companies reported that Black employees made up 14 percent of all employees, but dropped to 7 percent at the managerial level, and roughly 5 percent at the executive level. Black employees are leaving their jobs much faster than other races, citing *lack of career advancement* as a core challenge.

The study also found similar findings for Hispanic/Latino employees—they made up 11 percent of all employees, but only 8 percent of managerial jobs and 6 percent of executive-level positions. Both groups had a disproportionate number of workers in front-line, private sector jobs (19 and 18 percent, respectively), jobs that are low wage and have limited advancement opportunities.

Parallel to this, many workers said they would leave their employers for either not focusing on DEI or approaching DEI in a haphazard way.

[8] McKinsey & Company. (2022, June 27). Americans are embracing flexible work—and they want more of it. McKinsey & Company, from www.mckinsey.com/ industries/real-estate/our-insights/americans-are-embracing-flexible-work-and-they-want-more-of-it

After the 2020 elections, Gartner HR conducted a post-election survey in 2021 of 3,000 employees. Sixty-eight percent of respondents said they would think about leaving their current employer for one that takes a more vocal and visible stance on societal and cultural issues. Separately, in a 2022 well-being and voluntary benefits survey conducted by employee benefit consulting firm Buck, the findings showed that employees were likely to leave an employer if they think that the employer is not committed to DEI, diverse populations are not respected, or that their compensation and benefits packages are not responsive to the needs of diverse employees.

These points are especially important in retaining diverse talent. When recruiting for diverse talent, we have learned that we must meet talent where they are, focusing more on the unique skill sets and experiences that they bring to the table versus the school they attended or the degree they may or may not have attained. The same is true for retention.

A common mantra used in user experience design is "you are not your user." To that point, rather than letting our own cognitive biases guide our decision making on what an employee needs to thrive in the workplace, we need to holistically understand what *each employee needs* and adjust programs and offerings accordingly.

Let's now explore the impact of advancement opportunities and work flexibility on retaining diverse talent in finer detail, and how two companies—Target and Quartz—have pursued particularly innovative approaches.

Promoting Diverse, Mid-Career Talent

In our research we found an abundance of formal career advancement programs dedicated to advancing diverse, early career talent (professionals with less than three years of professional experience after high school or college) within private sector companies, along with metrics on candidate demographics and career outcomes. However, uncovering successful, formal programs designed to advance diverse *mid-career talent* (working professionals with more years of experience under their belt but still years away from retirement) to either leadership or other senior levels, proved challenging.

We're defining formal career advancement programs as those that have the following elements:

- A formal assessment of the worker's current skills, knowledge, and experience, as well as their career interests
- Specific and defined career goals, with targeted career opportunities (e.g., a program catered to management or executive leadership positions)
- A formal needs assessment of the skills and experiences that a worker needs to be ready for their target role(s)
- A clear plan, with timelines, of the training, tools, and resources that the worker should seek (training may or may not be tied to obtaining a formal degree or certification)
- A list of the people in or outside of the department/organization that will support the worker, through a combination of mentorship, coaching, and/or sponsorship
- Events and programming to solidify overall program objectives
- Periodic assessments of execution against the worker's career plan
- Periodic assessments of the program's effectiveness, along with participant demographic information
- Program costs are mostly or fully subsidized by the employer

We found many examples of such formal programs in academia and the nonprofit sector, and programs offered by external organizations, such as programs from Management Leadership for Tomorrow and the Executive Leadership Council to promote mid-career Black professionals to managerial positions. But we found few employer-grown, formal programs.

There are likely a few reasons behind this. First, employers may be engaging in these activities and practices, but not under the banner of a formal program. In many organizations, managers are expected to facilitate career conversations with their staff in creating an informal, individualized career plan, usually as part of their performance appraisal process. A frequent challenge with this approach is that it often places the onus on the employee to independently find the necessary experiences and skills to advance. Additionally, leaders who are newer to management, those who have poor management skills, or managers with numerous unconscious biases, may prove to be more of a hindrance than an advocate for a worker's career advancement.

WHAT ADVOCACY LOOKS LIKE

Sondra Sutton Phuong, an executive at Ford Motor Company who has risen through the ranks in the company over the past twenty years, recounted her own experiences when applying for her first supervisory role and how her manager made the positive difference in her initial career progression. "I was put on a 'perform first plan.' I got the job but was given it at a salary grade lower than the actual position. As I reflect on that situation now, I realize that my supervisor, Larry H. Collins, was the difference maker in that interview. He saw my potential and hired me when others on the panel had mixed views. After I was hired, Larry pulled me aside and said, 'Look, I am about to throw so much work at you that you won't be able to breathe. . . . But in 6 months, I will go back to the same team that would not promote you to supervisor and tell them to make it right.' He was a man of his word, and I will be forever thankful for his advocacy for me."

Second, formal mid-career advancement programs require a commitment of time, money, and administrative/operational overhead. For some companies, this is not feasible. For others, the expense involved doesn't outweigh a perceived risk—that trained employees will eventually leave the organization, and most likely, work for their competitors.

Third, for some employers, the development of such programs may go against their ingrained beliefs and perceptions of meritocracy, or that promotions and advancement opportunities should be given by merit alone, with "merit" often defined by subjective criteria. By developing career programs that target a specific, diverse audience, the concern is that underrepresented employee groups will be granted an "unfair" advantage, and not promoted based on abilities and contributions alone. Yet, it has been proven many times before that true meritocracy rarely exists, as personal biases cloud the decision for whom to give advancement opportunities or what "high potential" talent looks like, and exacerbates inequality across race, gender, and socioeconomic status.[9]

Additionally, diverse mid-career talent is likely in need of a more comprehensive and integrated program involving resources, mentorship, and sponsorship. In the McKinsey *Black experience* survey, mid-career Black respondents felt as if they did not have the access to tools or information to help them chart a successful career path. Who are the

---

[9]Makovits, D. (2020, January 9). How meritocracy worsens inequality-and makes even the rich miserable. Retrieved November 26, 2022, from https://insights.som .yale.edu/insights/how-meritocracy-worsens-inequality-and-makes- even-the-rich-miserable

people they need to reach out to? What is the nature of office dynamics and politics, and how can they navigate this to advance their career? If left under the guidance of a subpar manager, their success outcomes are likely to be lower.

Promoting from Within: Target Engineering Manager Immersion Program

Target has been cited not only as an employer dedicated to DEI in its internal and external endeavors, but also one that successfully retains and promotes top talent. With nearly 2,000 stores, Target is the seventh largest retailer in the United States. Founded in 1902, Target's headquarters are in Minneapolis, MN. The company also offers consumer financial services, through its debit, credit, and charge card offerings. Its total revenue in 2021 was $106B.

Target consistently has been lauded for its commitments to employees, sustainability, and the environment. The company is a certified *Great Place to Work*, is #13 on Fortune 100's *Best Companies to Work* for 2022, and was #1 in *People* Magazine's *Companies That Care* list for 2022. It also made the 2022 Diversity Inc. *Top 50* list.

Through a host of talent and leadership development programs, Target has been able to promote many internal candidates to senior management roles. Per workforce analytics company Revilio, the share of internal employee movement to senior management positions within the company was about 74 percent, out-pacing many other retailers, as well as financial services and tech firms.[10]

Target has made significant progress against its publicly announced DEI goals since 2019. In its March 2022 update on sustainability and environmental, social and governance (ESG) topics, the company reported:

- Promotions for people of color from exempt entry-level positions increased by 62 percent
- Turnover for staff members of color reduced by 33 percent
- Promotion of women to senior leadership increased by 16 percent

---

[10]Revilo Labs (2021, July 13). Which companies promote most from within? Revelio labs, from www.reveliolabs.com/news/macro/which-companies-promote-most-from-within

Tea Darden is just one shining example of Target's commitment to its employees' career trajectories. While in high school, Tea started as a team cashier for Target in 2003. Tea shared during the Great Place for All Summit in October 2022, "Leadership just saw the potential there and helped me to go through the motions." Tea progressed from a sales floor team member to an executive team lead in the course of nearly a 20-year career. Target's current chief diversity officer and senior vice president of HR Keira Fernandez also rose through the ranks, starting as a retail store team leader in Phoenix, Arizona in 2001.

Retaining and Promoting Women in Engineering

Even though Target has been consistently recognized by the Anita Borg Institute—a global organization dedicated to advancing women in technical professions and the host of the Grace Hopper Conference—as a *Top Company for Women Technologists*, like many other employers, it is challenged with not only finding female technical talent, but also challenged in advancing them into senior roles.

The Problem: Identifying and retaining women software engineers for senior-level technical positions.

Former senior director of technology Mitali Mathur noticed that many of her female mentees demurred when potentially facing an opportunity to make the leap into a senior-level role or leadership. Some of the concerns she heard were along the lines of "I don't think I'm ready" or "I still have a lot to learn," among others. Mitali herself felt the same way—"I had a similar mindset earlier in my career, so I know how intimidating the prospect of leadership can be. I wanted to help change that for others."[11]

Mathur, along with several other leaders and with talent acquisition developed the Engineering Manager Immersion Program (eMIP).[12] The program's mission is to increase diverse representation in technical leadership with a focus on underrepresented groups. The program was developed with the intention of building the critical capabilities so that candidates could be successful.

---

[11] Target Corporate. (2019, September). This tech program at target is inspiring and training diverse engineers for leadership roles.

[12] Mathur, M. (2019, October). How Target is Increasing Diversity in Engineering Leadership. Grace Hopper Conference 2019. Orlando.

The team started with a specific goal of increasing the global representation of female engineering managers by 2021. The program was also open to current Target team members and external candidates alike.

The immersive program lasts for 12 months and consists of four specific components:

- Formal learning focused on the leadership fundamentals
- Support for each participant, which includes an executive sponsor and a peer mentor, in addition to their manager
- Hands-on, on-the-job experiences to apply lessons learned from training
- Cohort development sessions, where other participants can share their experiences with one another, and have a sense of community

Upon completion of the program, participants are qualified to be considered for a senior engineering manager (SEM) role within one of Target's internal technology teams. The role comes with a considerable amount of leadership responsibility—specifically leading a team of engineers and making key technical and product development decisions.

Outcomes: Increased Representation of Women and Focus on Other Groups

Target launched the program in the fall of 2017, with 16 high-potential female engineering candidates from within Target and externally, geographically located from Minneapolis, Minnesota, and India. By the end of the first year, of the 16 candidates, 12 were accepted into SEM roles, roughly 75 percent of the initial cohort. As a result, there was nearly 10 percent more representation of women in SEM roles from 2018–2019.

During that time, Target took feedback from participants to adjust the program. For example, to ensure that mentors and sponsors could be more engaged with their charges, it adjusted the timing of the program to avoid peak workflow times that might hinder connection. For those external to the organization, the company began its onboarding efforts earlier, to account for the learning curve of getting acclimated to a larger, corporate environment.

Target also extended this program to other groups. Upon the success of the 2017 program, it extended its focus to include broader female representation (beyond those with technical backgrounds) in 2018. In 2019, the company continued to broaden its focus on Black and other ethnically underrepresented talent.

An alumna of the program, Anju Jha, who had worked as a lead engineer for seven years at Target prior to her promotion, articulated the impact the program had on her.

"After spending seven years at Target, I wanted to move into a management role and become an effective leader. This is when I decided to apply to eMIP. Thanks to the program, I now have more confidence and skills to handle a team, drive projects, and own innovation in our space."[13]

This is just one of many career and leadership development programs that Target invests in annually, specifically to support the growth of internal talent. Another is the *Dream to Be* program, which Target announced in 2021, in partnership with Guild Education. The program gives all part-time and full-time front-line Target employees the option to pursue select educational programs designed to help with completing preapproved high school, college prep, certifications, and formal higher education programs, all with 100 percent tuition coverage. Currently, there are over 250 course offerings in business management, IT operations, and design, to name a few.

Some could argue that Target's investment in training and development could have a negative return if it only served to prepare workers for better jobs outside the company. However, its investment has contributed to reducing turnover to a five-year low between 2019 and 2021, years when many retailers suffered from wide-scale departures.

Another potential benefit to recruiting with remote work in mind is that by no longer relying on geographic boundaries to source and recruit talent, employers can bolster their talent pools and pipelines considerably. This can be especially beneficial in less diverse areas of the country. In addition, hiring top candidates from lower-cost locations can reduce overall salary and benefits, although there are mixed reactions on the matter of variable pay based on an applicant's location.

The Positive Impact of Remote Work for DEI

Since the 1970s, remote work grew as the technologies that enable remote work improved, and more businesses realized benefits in cost reduction and employee productivity. For most employees, benefits include minimizing time and costs associated with commuting, more autonomy of how to balance their schedule against personal commitments, improved overall well-being, and being removed from political situations and gossip.

---

[13]Accelerating diverse technical leadership at Target. Target. (2022, May 31), from https://india.target.com/blog/accelerating-diverse-technical-leadership-at-target

It was not until the COVID-19 pandemic, however, that remote work became mainstream. For most businesses, it was crucial for them to continue to be operational, as well as to protect their employees as social distancing and facial mask mandates were put into place. In 2018, nearly 6 percent of the U.S. workforce did their work remotely; at the height of the pandemic in 2020, that increased to 36 percent. The current remote worker rate sits at 26 percent.[14]

As we emerged from the pandemic, employers began to ask their employees to return to the office. Even with the positive outcomes some companies have experienced, and with some admitting that critical work tasks are no harder to do remotely than they are in person,[15] criticism of remote work has included:

- Erosion of the *quality* of collaboration and communication
- Disincentivizing employees to perform to the fullest of their abilities
- Difficulty accurately assessing an employee's performance, from both a disciplinary and a career progression perspective
- Increasing an employee's feelings of isolation and invisibility

Kevin Walters, senior director of Diversity and Inclusion Solutions at SilkRoad Technologies, takes it further, saying, "it requires them to work harder. For example, managers will now have to track employees, schedule more meetings, find ways to engage employees, and find ways to effectively manage teams." Additionally, because of their existing commitments to "long term leases with office space," it is more of a driving factor to get people back in the office.

Some of the concerns raised are certainly valid. There is no doubt we have heard stories of workers abusing remote work. Stories have emerged of employees who have used work time to do everything other than their work, or in some extreme cases, where employees have worked two (or more) full-time, remote jobs simultaneously, leading to discussions on whether such actions are ethical and if this is a form of time theft (an employee accepting pay for work that has not been done).

But to borrow from Kevin Smith, chief product officer for software design company Abstract, "let's explore the connection between remote

---

[14]Flynn, J. (2022, October 26). 25 trending remote work statistics [2022]: Facts, trends, and projections, from www.zippia.com/advice/remote-work-statistics
[15]Merlini, K. P. (2022, November 17). Leaders Share Perspectives on managing remote workers. Retrieved November 27, 2022, from www.shrm.org/resourcesand-tools/hr-topics/behavioral-competencies/pages/leaders-share-perspectives-on-managing-remote-workers.aspx

culture, inclusion, and human performance." Despite these potential pitfalls, employers may be cutting themselves off from a variety of possibilities by demanding that all employees unilaterally return to the office, or at the very least, adopting a hybrid model, where employees come in for a portion of the workweek.

"Due to multiple socio-economic factors, diverse talent may experience several social challenges with access to transportation, childcare issues and other issues. A remote working arrangement allows diverse talent to be able to balance life and work accordingly," Walters says.

Current research tends to confirm Walters' view. Workers of color have expressed an overwhelming desire to keep remote work in place, primarily to avoid microaggressions and overt bias. SHRM conducted a study that found that 50 percent of Black workers and 29 percent of Hispanic/Latino workers prefer to work outside of the office.[16] "As long as there's an option to stay home, folks who are underrepresented are going to stay home because it minimizes their exposure to subtle acts of exclusion," says Tiffany Jana, founder of diversity consultancy firm TMI Consulting.

Women also have found remote work an important factor for their jobs. YouGov America discovered in its 2022 *Workforce Insights* survey that 44 percent of women surveyed cited that having a flexible working location is important, for many of the shared reasons of Black and Latino workers. The flexibility also allows women to continue to build their careers while caring for their families and household responsibilities, as they are still more likely to assume the lion's share of household management tasks.

Finally, remote work has made employment possible and more accessible for workers with disabilities. In fact, those with disabilities have championed for the longest to receive accommodations to work from home; otherwise, depending on the nature and severity of their disability, having gainful employment may be almost impossible. While 1 in 4 adults in the U.S. identify as having a disability, only 19.3 percent of people with disabilities were employed in 2019.[17] While disability representation *decreased* from 2019–2021 during the pandemic, this decrease can

---

[16]Janin, A. (2022, May 15). Some minority workers, tired of workplace slights, say they prefer staying remote, from www.wsj.com/articles/minority-workers-prefer-remote-microaggresions-11652452865

[17]National Council on Disability. (2021, May 3). 2020 progress report on national disability policy: Increasing disability employment. Retrieved November 2022, from https://ncd.gov/progressreport/2020/2020-progress-report

be attributed to the higher tendency for disabled workers to be in positions more susceptible to furloughs and layoffs than their non-disabled counterparts.[18]

IMPROVED DIVERSITY THROUGH REMOTE WORK: QUARTZ

One company that has utilized remote work successfully is the online platform, Quartz. Established in 2012, Quartz is an online business and technology news platform, founded by former staffers from *The New York Times*, *The Economist*, and the *Wall Street Journal*. The platform's website, qz.com, averages slightly over 3 million global visitors to its site monthly; 58 percent of which reside in the U.S.[19] Like many companies in 2020, Quartz evaluated improvements to its recruiting practices to become more inclusive. And as with other companies, a tight labor market also made it more challenging to attract talent generally. One of the company's new approaches included recruiting for positions anywhere in the world, rather than concentrating only on cities known for being magnets for publishing talent, like New York City.

What resulted was a rapid increase in not only the quality of candidates, but the diversity of the candidate pool as well, and much faster than expected—within the time frame of one year. "Nothing compared to how fast and strong an impact opening up our applicant pool to remote applicants had," said Quartz CEO Zach Seward. "The quality and diversity of applicant pools [for] most positions increased dramatically," Seward said. In 2021, 42 percent of the staff were comprised of people of color, up 12 percent from the year prior.[20] Quartz applied the innovation principles of Courage, Risk-Taking, and Collaboration with its adoption of remote work and has reaped the benefits from it.

What's Needed for Hybrid and Remote Work Success

Despite the challenges and risks that managers can face when it comes to implementing hybrid and remote work, it can be a wonderful and beneficial modality for employers to embrace and implement at scale.

---

[18]U.S. Bureau of Labor Statistics. (2022, February 24). Persons with a disability: Labor force characteristics summary—2021 A01 results, from www.bls.gov/news .release/disabl.nr0.htm

[19]Semrush. (2022, December 19). Domain overview - qz.com. Retrieved December 2022, from www.semrush.com/analytics/overview/?q=qz.com& searchType=domain

[20]Guaglione, S. (2021, November 17). Why publishers say opening up remote hiring has grown and greatly improved the applicant pool, from https://digiday.com/ media/why-publishers-say-opening-up-remote-hiring-has-grown-and-greatly-improved-the-applicant-pool

Executive Managing Director and Global Future of Work Leader at Jones Lang Lasalle (JLL), Peter Miscovich, believes that leaders' own biases often can get in their way of successful hybrid work and remote work adoption. Citing research from McKinsey, a contingent of managers over the last 30 years who may have proximity bias (preferential treatment to those physically around us), adjacency bias, and cultural bias—as these biased management behaviors are often "difficult to let go of" and these biases will impede successful hybrid work program adoption.

In addition to addressing these challenging proximity and cultural biases, Miscovich believes that the following elements must be in place for hybrid and remote work programs to be successful:

- **Empathy from leaders**—"If you have true empathy for your workers and your workforce, you should be able to start to understand—what does a microaggression look like? Why might someone of a certain demographic be uncomfortable in certain workplace settings? With empathic leadership and management practices—progressive leaders can fully engage and understand how individual employees may be feeling and if these employees are uncomfortable with certain workplace dynamics, including microaggressions. Per Miscovich, by understanding the needs of individual employees fully—organizations can better enable and support empathic leadership and management skills that will provide the optimum high-performance workplace for their employees."

- **Trust**—While there may be challenges regarding trust in employees performing their work remotely, "I think within specific industry sectors, those abuses of trust may tend to occur with some frequency, and perhaps there are sectors that require special provisioning and oversight as a result. But in general, most people want to do a good job, most people want to be trusted. Most people want to perform well for their manager and for their organization." Miscovich advises against creating remote work policies and invasive "surveillance" practices that may negatively affect a broad audience, but are only addressing a small portion of the employee population who may "behave badly". Trust is an essential element for any hybrid work program to be successful for the long-term.

- **Creating immersive employee experiences**—Even in hybrid and remote work, the need to create immersive employee experiences with social elements will still be required. "We need more intentionality around immersive social engagement—How can we curate that level of immersive social engagement? How can we create the immersive and experiential employee engagement beyond the traditional physical office environment?" Experiences should be developed in an intentional way, with the aim of helping promote greater inclusivity and belonging. The digital and physical workplace ecosystems that are emerging must be fully connected and highly technology enabled to provide the most desired "human-centric" and immersive employee experiences. The digital workplace and the physical workplace must both create a strong employee magnetism that will draw employees together—to engage more fully with one another with greater commitment, inclusivity and belonging.

- **Performance management systems**—a solid performance management system allows leaders to track employee performance in a clear, consistent, and measurable way. Performance management systems consist of a blend of work processes, HR methodologies and technologies; ideal performance management systems work on a model of continual improvement and innovation over time and have employees actively participate and collaborate with their managers with their career development.

- **Harmonization on a programmatic level**—Miscovich says, "Leaders and HR partners need to be in lockstep as to how remote and hybrid work policies and processes are developed and executed. While there is no 'one size fits all' approach—the approaches that work for company A may not work well for company B. A programmatic hybrid workplace program framework needs to be established and agreed upon by all key stakeholders with the ability to refresh and update as an 'evergreen' hybrid work program framework over time. This 'harmonization' of the program needs to work effectively for everyone—for executive leadership, managers, employees and HR partners to successfully engage, enable and implement hybrid work programs over time. Team leaders should be able to differentiate hybrid work practices for their unique team requirements and individual team member work behaviors. Teams within an organization will be able to 'harmonize' their hybrid work behaviors as may be required to meet key financial outcomes and business performance. Hybrid work program harmonization over time should serve to meet the key business performance outcomes and financial objectives versus a company-wide blanket hybrid workplace policy that is set for the entire organization."

- **Top notch technology**—"If an organization is going to implement hybrid or remote work practices, the organization must provide robust technology enablement," says Miscovich. "Organizations need to provide consistently good fidelity of both video and audio." Additionally, as mentioned in JLL's October 2022 research report, "Technology and innovation in the Hybrid Age", technology adoption goes beyond making one-off technology investments that are then shoehorned to fit a current workflow. Rather, for hybrid work programs to be successful, organizations should think carefully about proactively designing integrated technology systems with their changing and evolving "hybrid" work processes.

- **Flexibility and room to evolve**—"We question the viability of the three- day, four-day, or even two day [work in the office] mandates— company-wide office work mandates across an entire organization generally don't really work for the long-term as sustainable hybrid work policies," Miscovich says. Instead, the development of flexible hybrid workplace guidelines that are fair to the employee, manager and to the organization are more sustainable and much more effective. Additionally, hybrid and remote work programs should be able to change over time due to the evolving and changing needs of individual employees and teams. The elongation of the COVID-19 pandemic certainly frustrated many managers, but the elimination of hybrid work or remote work completely isn't the answer. "If we should experience another [COVID-19] variant in the future that becomes problematic, then we will have to allow for the workforce behavior to shift once again towards remote work to accommodate that new variant threat. We must be able to accommodate and adapt to whatever new workplace threats may emerge over time—and we should consider hybrid work and remote work as essential workplace strategic solutions for long-term organizational flexibility and resiliency."

Compensation: Innovate with Employee Benefits Programs

Employers in the U.S. are not required to provide comprehensive benefits to employees. Right now, there are only three basic benefits that all employers with at least one employee are required to give by federal law: unemployment, workers compensation, and Social Security. But many know that, in conjunction with a competitive salary offer, benefits packages that include quality medical insurance, paid time off, and retirement plans among others, are essential to competitively attracting and retaining top talent.

There's a disconnect, however, between what employers *perceive* as a great benefits package and what employees *believe* constitutes a great benefits package. The disconnect is even larger between employers and diverse talent. With overall compensation reported as being the number one reason employees leave their employer, it's important to narrow or eliminate the perception gap. The previously mentioned well-being and voluntary benefits survey by Buck illustrates the differences on the voluntary benefits employers and employees value (see Figure 8.1).

Voluntary benefits employers value

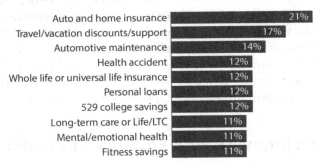

Voluntary benefits employees value

Figure 8.1: Differences between benefits employers and employees' value, 2022 Well-being and Voluntary Benefits Survey

Reprinted with permission from Buck, LLC

While many would agree that assistance with caregiving can be extremely beneficial, we can see that employees benefits that will allow them the opportunity to protect or improve their overall finances, and save money on everyday expenses, were preferred. Having benefits that also support employee mental health and wellness were also important. This is a prime opportunity for employers to think outside the box and apply innovation principles in meeting and supporting employees where they are and where they strive to be.

The focus on financial health and well-being is perhaps not surprising, with the ongoing uncertainty of economic conditions in the country and the still lingering economic impacts of the COVID-19 pandemic for certain populations. Bank of America highlighted in its 12th annual *Workforce Benefits* nationwide survey of 824 employees and 846 employers, that due to financial difficulties, half of employee respondents had to either tap into their emergency savings (21 percent), work overtime (21 percent), find higher paying employment (20 percent), or take a hardship withdrawal from their 401K retirement plan (6 percent).

People of color and women are particularly vulnerable to negative financial impacts. They are already at a particular disadvantage as they have been historically underpaid versus their white, male counterparts. They are also likely to be starting their careers with an already heavy financial burden than other groups, in the form of student loans, and less likely to be able to build generational wealth. Women held 58 percent of all student loan debt and nearly 90 percent of Black graduates have taken out student loans. Both groups have taken on a sizable amount of debt to complete their studies and are still repaying this debt well beyond 10 years.

But given that employers are trying to cater to a host of different audiences, it is important for them to take the time to understand the benefits that would be the most impactful and meaningful. There is no one size fits all—gender, race, ethnicity, and age are just some of the ways that will impact the mix of benefits offerings.

For example, offering a cash plan where employees can pay for a variety of everyday costs and expenses can be cost effective and appeal to a wide variety of employees. During the Great Place to Work For All Summit in October 2022, Americas & US Diversity, Equity & Inclusiveness Leader Leslie Patterson shared highlights of Ernst & Young's (EY) Way of Work initiative. Aimed at employees coming back to work full-time, they launched a pilot where every EY employee would receive $800 to use at their discretion—commuting costs, pet sitting, childcare, etc. To fund the program, they redirected funds from underutilized or unused benefits. They plan to roll the program out again to see if and how funds from the plan are being utilized.

Student debt repayment plans (SLRPs) can also be an attractive incentive for workers. Employers, usually through a third-party service, can send their contributions to an employee's student loan company. No doubt this can be a costly benefit to both parties. SLRPs do not currently have tax-free status, so employers cannot get a tax deduction, and any payments are considered taxable income. With this in mind, it can be a strong incentive for recruiting Millennials and people of color. In 2017, Aetna (a wholly owned subsidiary of CVS Health, #28 on DiversityInc's Top Companies for Diversity) launched a SLRP program, where it matched employees' student loan payments for full-time employees up to a lifetime maximum of $10,000. If too costly an option, employers can instead partner with outside organizations in offering reduced student loan interest rates when refinancing.

With overall compensation the leading reason employees leave their companies, innovating to align benefits to the needs of employees can have a profound impact on retention.

Table 8.1 reviews the approaches Target took to retain and grow its diverse talent, while Table 8.2 reviews how it minimized potential innovation threats.

Table 8.1: Aligning Retention Practices with Innovation Principles

PRINCIPLE NECESSARY FOR A CULTURE OF INNOVATION	PRINCIPLE IN ACTION
Risk-Taking	▪ While Target boasts many talent and leadership development programs, not many are as long in duration or as comprehensive as the eMIP program. There was a risk not only with having participants ready for substantial leadership roles, but to have them remain in their roles, advance, and succeed.
Collaboration	▪ Working across many functional areas in Target, the eMIP leaders worked to communicate the program to broader audiences like social media, and through prominent platforms, such as the Anita Borg Institute.
Leadership	▪ Senior leadership took an active, hands-on role throughout each cohort, to ensure that participants were ready to assume leadership roles.

Table 8.2: Innovation Threats and Preventative Actions

INNOVATION THREATS	ACTIONS TO AVOID THREAT
Inertia 	▪ Not ones to rest on the pilot's success, the program's leadership took feedback and revamped the program to be more inclusive to broader audiences.

Conclusion

Recruiting and onboarding diverse talent is only part of the equation. For diverse talent to be successful and stay within organizations, companies must couple leadership and culture with tangible tools and resources. As explained in Chapter 2, equity means providing employees what they personally need to be successful. Employers can leverage career advancement programs, remote work, and benefits programs to give diverse talent their best chances for success and, at the same time, reduce costly turnover.

Summary

- Employee retention and employee turnover are important data points to assess hiring practices and management effectiveness. It can also provide insight into the health of the culture of the organization.

- Lack of compensation, workplace flexibility, and career advancement opportunities are the primary reason people leave their jobs.

- Since the pandemic, women and people of color have experienced the most turnover in positions at all levels and across industry sectors.

- Career advancement programs can help propel mid-career talent within the organization and improve overall retention.

- Remote work not only is a tool to improve retention, but can help diversify (and improve) a company's total applicant pool.

- Understanding the needs of diverse work populations can help companies craft unique, more highly valued, and cost-effective benefits programs.

The Inescapable, Undeniable Role of Executive Leaders

Leadership is one of the five organizational characteristics that create conditions for innovation, regardless of industry or function, that we introduced in Chapter 3.

In Chapters 4–8, we met several mid-level leaders and small business owners who took reputational and professional risks, secured senior-level support, and built coalitions to solve root problems of talent acquisition, development, growth, and retention by *innovating for diversity*.

For executives leading large, complex enterprises, the stakes are even higher. Senior-level leadership is especially vital in expanding an organization's capacity to innovate for diversity. We devote this chapter to exploring how effective senior leaders, including CEOs, promote cultures that prioritize DEI principles as business imperatives. Using the examples of the National Hockey League and World Wide Technology, we will examine the specific priorities and actions of their C-suite executives who have successfully pioneered new DEI approaches that have transformed culture and improved business outcomes.

Key Concept: Powerful proclamations of a commitment to diversity by CEOs can be motivating to employees and other stakeholders, but sustained support for organizational systems and establishing measures and accountability is required for DEI principles to be truly embedded in the culture of any organization.

Words and Actions: Do Behaviors Match the Script?

Throughout 2020 in response to the murder of George Floyd and the Black Lives Matter movement, CEOs of companies large and small announced ambitious commitments to improve their DEI practices and outcomes. As of July 2022, for the first time, all Fortune 100 companies have made a public commitment to diversity, equity, and inclusion with their plans summarized on their websites. In the same time frame, over 2,200 U.S. CEOs have signed the pledge promoted by PwC's CEO Action for Diversity and Inclusion committing to "support a more inclusive workplace for employees, communities, and society at large." And, according to a 2021 survey by the advisory firm Glass Lewis & Co., a quarter of U.S. companies included some form of environmental or social metric—including DEI targets—as part of their executive incentive plans, up from 16 percent in 2019. A June 2022 Fortune/Deloitte survey of 116 CEOs representing over 15 industry sectors found that 92 percent of respondents acknowledged "DEI has been built into my strategic priorities/goals as CEO," up from 61 percent reported the prior year. Of note, 71 percent reported that their board expected regular reports on their progress against DEI goals, compared to just 49 percent the prior year.

These actions point to the role of the CEO as a personally engaged and visible leader in shaping and communicating DEI strategies. CEO commitment to diversity both publicly and within the organization is widely understood to be table stakes for successfully building an inclusive culture. Michael Bush, CEO of Great Place to Work, takes this a step further, "CEOs must be unwavering in their commitment to DEI; otherwise, staff will see efforts as half-hearted, and you just won't see progress." He compares pervasively executed DEI practices with the implementation of enterprise-wide customer relationship management software, "The CEO wouldn't say, 'we've just invested tens or hundreds of millions of dollars in a new software platform, 'use it, if you want,' or 'try it and see how you like it.'" Commitment to DEI must be taken as seriously as any significant strategic or systems change within a company.

A CEO's words and actions alone are not enough for DEI to take root, but without them the rest of the organization won't follow. A 2018 study by Dr. Eddy S. Ng and Dr. Greg J. Sears, "Walking the Talk on Diversity," examined the critical link between a CEO's earnest pro-diversity behaviors and the efficacy with which HR managers and middle managers implement diversity practices. CEOs who take a resource-based view see diversity as an opportunity to enhance creativity and performance while those who are wary of DEI practices as another vehicle for stereotyping tend to see greater diversity as a source of inter-group conflict and a threat to organizational effectiveness. Those CEOs with a resource-based perspective more successfully achieve organizational buy-in over the long term as well as a culture where managers are comfortable prioritizing diversity management over other competing demands.

This research is consistent with findings from a study published in Harvard Business Review, "CEOs Explain How They Gender-Balanced Their Boards," by Dr. Stefanie K. Johnson and Kimberly Davis (whom we will return to later in this chapter). Johnson and Davis interviewed CEOs from 36 companies in the S&P500, 15 of whom had achieved gender parity on their boards. Their conversations revealed how CEOs who successfully influenced the recruitment of more women to their boards were more likely to have a strong "promotion" focus when pursuing goals. That is, they were motivated by what they stood to gain, their aspirations, and a desire for improvement. Other CEOs tended to have a strong "prevention" focus, motivated to protect what they might lose.

John "JT" Saunders, chief diversity officer at Korn Ferry, sums up the CEOs with a strong promotion focus and resource-based view of diversity as those who are most likely to lead with courage and tenacity. Before he became chief diversity officer, JT served in a role at Korn Ferry where he advised CEO clients on organizational design and DEI principles. From his experience he has concluded, "Courageous leaders are not only champions, but they inspire change and hold people accountable. They don't let the operationalization of DEI get gummed up in middle management, but ensure all managers have the understanding, awareness, tools, and clarity for what they are responsible for." JT shared his observations of a number of well-intentioned CEOs superb at externally signaling their beliefs and commitment to a diverse workplace, but who fail at embedding necessary processes internally, "If DEI is viewed as something we do in one corner of the organization and not implemented across a company with the right resources and training, then it will become the first thing to backslide when the business cycle dips down."

A Focus on Systems Thinking

The type of comprehensive—and operationalized—approach to DEI that JT Saunders describes requires the same sophisticated systems-level thinking needed to manage any large, complex organization. Systems-level thinking requires leaders to understand not just the individual parts of any process, but how those parts interact with one another. In fact, a focus on systems thinking surfaced as a consistent theme in many of our executive interviews.

Sondra Sutton-Phung, the executive at Ford Motor Company whom we met in Chapter 8, talks about the need for DEI to be a systems-level process refined over time, "Leaders can sometimes consider DEI a project with a start and end date when it's actually a continuous process." Among the many "parts of the process" she suggests any company critically examine are pay equity, employee training, recruitment and advancement, mentorship and advocacy, and manager accountability, not as one-off projects, but together as a holistic system with clear goals.

Our premise of the mutually reinforcing benefits of innovation and diversity—diverse teams boost innovation and in turn innovation improves DEI practices and outcomes—can also be applied to systems thinking.

Stephen Ezeji-Okoye has tackled complex systems his entire career. He transformed the delivery of urgent patient care and rehabilitation services while serving at the U.S. Department of Veterans Affairs, innovated public-private partnerships to improve access to quality healthcare for veterans, and today is re-inventing affordable primary healthcare services in his role as chief medical officer for Crossover Health. Stephen believes that none of these massive, complex systems challenges could be rewired without an abundance of diverse thought, experience, and perspectives, "For systems to change, we need people who can view the world as it could be, not what it is today. That requires diverse opinions willing to challenge the status quo, and who can look at a problem from an outsider's point of view." Stephen talks about feeling like an outsider as a child and through his early adult years. He was born in England to a Nigerian father and an English mother and spent his formative years living in and navigating the disparate cultures of England, Nigeria, and Canada before moving to the U.S. for college. In his own career he has prioritized hiring diverse thinkers who, like him, bring fearlessness to systems-level problem solving.

Here we see the idea of another virtuous cycle in action—greater diversity advances systems change, while systems change is often required to address root causes preventing diversity at organizations to thrive.

Let's take a closer look at how, and importantly *why*, senior leaders at two very different organizations—the National Hockey League and World Wide Technology—are embedding DEI practices throughout their business operations and taking a comprehensive approach to underlying systems change. Leaders at both organizations are keenly focused on building cultures of inclusivity to foster innovation that delivers greater value to all stakeholders, including employees. They have intentionally recruited and developed senior teams with a promotion-focused, resource-based view of diversity and who act at a systems level to influence longer-term goals.

National Hockey League: DEI as a Movement, Not a Moment

Kim Davis joined the National Hockey League (NHL) as executive vice president in 2017, the year the League celebrated its 100th anniversary. For the ten years prior, 2006–2016, the fan base had aged more than that of any other major sport. According to a 2017 study by McKinsey, fewer millennials and GenX'rs identified as "committed fans" of the NHL, than of the National Basketball Association (NBA), National Football League (NFL), or Major League Baseball (MLB). At the same time, the fan base for hockey remained overwhelmingly white at 84 percent. Given the changing U.S. demographics summarized in Chapter 1, it doesn't take a seasoned demographer to understand that without intentional change, the future of hockey was in question.

And this is where our story begins of how Kim, working closely with NHL commissioner Gary Bettman and colleagues across the League, fired up a movement and, in the process, *innovated for diversity*.

Before joining the NHL, Kim had a successful career in banking, philanthropy, and strategy consulting. She progressed through the ranks at JPMorgan Chase over a 20-year tenure and became the first Black woman promoted to a senior vice president role. She ascended to become president of the JPMorgan Chase Foundation, which she ran for over seven years before joining the global advisory services firm, Teneo.

Indeed, her unique background in finance, community building, organizational design, and civic engagement proved to be a powerful mix of experiences when the NHL became her client at Teneo. That she was a Black woman who could view fresh opportunities for the League through her unique experiential lens was but one factor in helping shape a new future for the sport.

To advance the best interests of her client, Kim conducted deep research into the attitudes and composition of the League's fan base, as well as the perceptions of those not committed to hockey. After analyzing the data, her central recommendation to the Commissioner was a comprehensive, long-term plan to grow the sport with a diversity-anchored strategy. After the Commissioner reflected on the findings, he asked Kim to join his team as a direct report in a new role to increase fandom, value creation, and growth through increased youth participation, social impact, and outreach to new and underrepresented audiences. Interestingly, when Kim was asked to consider the framing of her new title, she insisted that any reference to "DEI" be left out. As senior executive vice president for social impact, growth, and legislative affairs, much of what she does involves building greater diversity and inclusion into the sport at all levels—*as a business*.

Kim speaks forcefully and with passion about the imperative for DEI practices to be "woven into the very fabric of the business, into the DNA." In our interview she observed how over the course of her career she witnessed the creation of numerous diversity programs that quickly dissipated when resources became scarce, or champions left for other roles. But, when DEI is established as integral to business success, it also becomes integral to the culture.

In her new role, Kim identified seven "dimensions of excellence" required to achieve the seamless integration of culture and diversity that would lead to long-term business success as the League entered its second century: Leadership, Education, Employment, Marketing, Partnerships, Community & Civic Engagement, and Youth Participation. With the full support of the commissioner and input from key executive colleagues, she plotted a framework of tactics and success measures for each of the seven dimensions. We'll take a closer look in the next sections at two: Employment and Youth Participation.

Employment: Building an NHL Talent Strategy to Reflect Local Communities

Within the first 10 minutes of just about any conversation with Kim Davis, one quickly learns she does not jump into a new endeavor without first evaluating the data and the facts. Not surprisingly, she wanted a baseline to understand the composition of the internal workforce at both the League and team levels before she could advise on strategies to attract diverse talent. It was widely known that over 90 percent of professional players across the 31 teams (as of October 2022 there were 32 teams with the expansion of the Seattle Kraken who began play during the 2021–22 season) were white, but staff composition within scouting, operations, and front office roles was less clear. She soon realized there was no baseline. No comprehensive review of staff composition existed.

When Kim shared that piece of information with us, we weren't terribly surprised, nor was she. In our interviews with numerous executives across a range of industries, many revealed that their employee diversity tracking extended only to recruitment of new hires and did not measure the full picture of how employees advanced through the organization, nor did they break down gender, race, ethnicity, sexual orientation, languages spoken, veteran status, educational background, or ability by occupational level. The insight here is that DEI will never be prioritized if it is not vigilantly monitored and measured like any other business deliverable, such as sales figures, market share, or brand awareness. If it's not measured, it doesn't matter.

Once Kim completed a baseline analysis, she was not surprised to find significant underrepresentation of women and non-white employees across the organization. As uncomfortable as it may have been to confront the facts, it was precisely the starting data required to begin a creative, intentional strategy to recruit and promote differently. Efforts to reach into different talent pools paid off at both NHL headquarters and in scouting and operations at the team level.

In 2021, Kim identified a range of new talent sources, including the diversity and inclusion technology company, Jopwell, from which to recruit underrepresented talent for league headquarter roles. Within just nine months of intentional effort, the number of women and multicultural staff members jumped 36 percent. At the same time, she advanced opportunities for women in hockey operations where only one woman had ever been hired in the entire history of the NHL. Professional women's hockey had never been a source of scouts or coaches but proved an excellent source of such operational talent. At Kim's suggestion, the NHL launched a Player Inclusion Committee comprised of team owners and executives, which included eight leaders from women's hockey. Within nine months, all eight women were recruited to join NHL teams in operational roles. From January 2022 to October 2022, 10 women were working in such roles, a dramatic uptick from the baseline of "one" ever hired. Kim noted, "What these talented women needed more than anything was exposure to club executives making hiring decisions—that's what it took."

The NHL Coaches' Association (NHLCA) contributed to the goal of diversifying the sport by launching several initiatives in 2020–2022 to recruit from the National Collegiate Athletic Association (NCAA) and youth networks, as well as from international programs, to develop player (and coach) roster compositions more reflective of the communities hosting NHL teams. Two initiatives included the NHLCA BIPOC Coaches Program and the NHLCA Female Coaches Development Program with the goal to support female and minority coaches with the skills development, leadership strategies, and networking opportunities to advance their careers.

There is also a growing interest among wealthy Black and Latino leaders to invest in the sport. As of 2022 former NFL running back Marshawn Lynch, entertainers Will Smith and Jada Pinkett Smith, film producer James Lassiter, Black Entertainment Television co-founder Sheila Johnson, investor Earl W. Stafford, and real estate and media entrepreneur Alex Meruelo were all either minority or majority stakeholders in NHL teams. Dave Stewart, founder and chairman of World Wide Technology, who we profile later in this chapter, holds a minority stake in the St. Louis Blues. Many of these influential leaders also lend their voice and their networks to recruiting diverse professionals to the sport, as players, announcers, coaches, or administrative executives.

Youth Participation: Diversity as a Growth Mindset

Any major league sport requires a healthy, growing fan base to thrive. We learned that a person is 3–5 times more likely to become a fan of professional ice hockey if they played it themselves, or if one of their children or relatives played the game. Yet, hockey is not the most accessible sport to play. Equipment is expensive and ice rinks are rare in many rural towns and urban hubs. And, in communities where there is little tradition of hockey, the game lacks cultural relevance. Expanding the geographical and demographic reach of ice hockey requires a very different approach than baseball or basketball.

In a number of traditional NHL markets, fans fell in love with the sport through street hockey. Playing with skates or sneakers, a stick, and a small ball teaches the same foundational principles of the game, without the need to be on the ice. To build enthusiasm for street hockey and a future NHL fan base, Kim formed a partnership with RCX Sports to expand street hockey to 5- to 17-year-old boys and girls in markets across the U.S., with an emphasis on regions with no professional hockey team such as Albuquerque, NM, and Austin, TX. Beginning in 2022, RCX Sports is organizing youth leagues, tournaments, and source equipment and uniforms with the objective of reaching hundreds of thousands of soon-to-be hockey enthusiasts.

The partnership with RCX Sports demonstrates both cleverness and humility. "How do we build trust in markets with no NHL presence and with audiences who don't see themselves in the sport?", Kim asked when evaluating strategies to grow the fan base. That question was in part answered with the structuring of the RCX Sports partnership. For years, RCX had built good will and trust in multiple U.S. markets through their NFL FLAG program with over 500,000 girls and boys ages 5–17 playing non-contact flag football across 1,600 locally operated leagues. Now, a new generation will be far more likely to identify as hockey fans, thanks to an approach that viewed geographical, racial, and gender diversity as a growth strategy.

Systems as a Movement-Making Innovation

How Kim and the NHL are tackling Employment and Youth Engagement are just two of the seven *dimensions of excellence* identified as necessary for the culture to support and value greater diversity at all levels. And, in turn, to deliver exceptional business results. During the 2021–22 season the NHL achieved record revenues, in excess of $5.3 billion. And while there are certainly many reasons for that success, the steady and persistent efforts to fine tune the seven dimensions of excellence contribute to the long-term health of the league.

And it is the long term that Kim continually keeps in her sights. When asked about her own legacy at the NHL, she is quick to respond that she expects to establish a set of systems and accountabilities for culture so embedded in the business that a change of leadership would not abandon them, but rather evolve them. Elaborating, she noted, "When a new leader comes on board, we don't expect them to abandon sponsorships, player development, or operations; these are fundamental aspects to the business of the NHL. Likewise, I would expect our focus on demographics and diversity as a growth strategy to be similarly embedded as part of how we conduct our business long into the future."

At the NHL, through Kim's leadership and the support of Commissioner Bettman, the very practice of interweaving "diversity as a growth mindset" across the core business through the seven dimensions of excellence, evolving culture, and reinforcing new practices with systems of accountability all represent innovation. The understanding that building a sport accessible and enjoyed by all truly takes *a movement, and not a moment* and requires that sustained commitment and discipline is at the root of the NHL's *innovation for diversity*.

World Wide Technology: The Power of Successfully Scaling Culture

World Wide Technology (WWT) is a $14.5 billion global technology services company founded in 1990. Headquartered in St. Louis, MO, it consistently ranks among the 20 largest privately held companies in the U.S. WWT is regularly featured on industry lists—Fortune, Great Places to Work, Time, Glassdoor—as a top employer and an especially welcoming workplace for women and minorities underrepresented in tech. After the early years of its founding, the company has consistently grown year-over-year revenue, with the exception of a small dip during the 2008 recession, largely through organic growth, not acquisition.

A deep commitment to an unwavering set of values undergirds the foundation of the company's success. Unwavering is important in any discussion of WWT and culture. Unique to most tech companies, the two co-founders remain active some 30 years after starting the business and continue to propagate the same powerful belief system that shaped the earliest days of the company. The fact that they have effectively scaled their culture with now nearly 9,000 employees in over 15 countries is a tribute to their singular, sustained focus on a set of principles formulated by the founders early on.

The Story of WWT's Culture and Values Begins with its Founders

Perhaps not surprising, WWT's culture is inseparable from the character of its founder and chairman, Dave Steward, and CEO Jim Kavanaugh. Both have their roots in the Midwest and both share an appreciation for determination and resilience, prioritize trust, teamwork, and fair play. They embrace a similar set of values forged from very different life experiences.

Dave Steward grew up nearly four hours from St. Louis on a farm in rural Clinton, Missouri. One of eight siblings, he speaks affectionately about the love and encouragement of his parents, and the grueling work of early morning farm chores. His father was Dave's model of an entrepreneur, at times a mechanic, bartender, and trash collector, and owner of a janitorial service. The support of his close-knit family and a deeply rooted faith buttressed Dave and his siblings against the profound challenges of being among the first Black families to integrate into local schools, movie theaters, and public swimming pools in the 1950s and 1960s. In 1969, Dave was accepted into Central Missouri University and won a basketball scholarship his sophomore year after his coach took note of his relentless work ethic.

Jim Kavanaugh was quite certain he would work in construction or start a landscaping business after high school. Growing up in St. Louis with a bricklayer father and homemaker mother, college was not a forgone conclusion. Hard work, the grit to push through challenging times, and a generally optimistic attitude were valued qualities throughout Jim's upbringing. A gifted athlete who trained tirelessly, Jim won a scholarship to play soccer at St. Louis University and later earned a spot on the 1984 Olympic team and the Pan American team. After completing his degree and with just $25 in his pocket, Jim moved to Los Angeles to play soccer professionally with the Lazers. His dedication to teamwork and the discipline required to constantly improve—together with a healthy competitive spirit—served him well on the field and in his first tech sector job.

The story of how Dave and Jim formed a business partnership over 30 years ago reveals the role that trust and integrity plays in their relationship and in WWT's corporate culture.

After a wildly successful decade-long sales career at Union Pacific and Federal Express, Dave acquired his first company, Transportation Business Specialists, that audited shipping invoices for several Fortune 500 companies. When the business expanded, he landed a large contract with his former employer, Union Pacific, that required sifting through three years and $15 billion worth of bills. He built from scratch what he believed at the time was the largest local area network in St. Louis to secure the computing resources needed for the task. From that moment he knew the future of his entrepreneurial endeavors would be anchored in technology.

At the same time, after leaving professional soccer, Jim was selling circuit board components for the electronics distributor, Future Electronics. His boss was contemplating starting his own electronics distribution company and enlisted Jim to join him. Jim's boss also knew Dave Steward and convinced Dave not only to come aboard, but to invest much-needed start-up capital. The new venture was off to a successful start, but after a year Jim began to doubt the integrity of his former boss's business practices. Believing he could trust Dave and following his gut, he found the courage to oust the former boss.

From that moment, the foundation was laid for Jim and Dave to grow World Wide Technology, deliberately guided by a clear set of shared values.

Bob Ferrell, WWT's chief diversity officer and executive vice president, observes that the importance of diversity at the company is not new; WWT's diversity ethos originates from the long-term partnership, friendship, and mutual belief in how business should be conducted between the two founders, a Black man and a white man. Diversity is encoded at WWT as the first among four corporate goals:

1. World-class culture built on diversity and inclusion
2. World-class execution
3. Customer delight
4. Double every five years

Codifying and Scaling Culture at WWT

It's easy to imagine ways any tremendously successful technology service company grows through effective deployment of technical innovation. Indeed, WWT has pioneered new approaches to testing and evaluating technology solutions for its customers at their Advanced Technology Center (ATC). The ATC is designed as a collection of physical labs, virtualized to create a collaborative environment for clients and WWT technologists to design, build, and adopt technology products and systems.

What's harder to imagine is that arguably the most important innovation at this tech company is its investment in and methodology for scaling culture, which in turn has led to further insights and innovation specifically for advancing diversity and inclusion, as well as in producing technology breakthroughs.

Let's start with the underlying principles that shape WWT's culture. Ten years after the company's founding, Jim and Dave landed on the importance of codifying core values, eight key business concepts, and a consistent framework for developing employees that assesses in equal measure job performance *and* demonstration of values. Jim personally architected what is now known as Integrated Management and Leadership (IML) that is foundational training for every new employee, reviewed annually as part of company-wide offsites with all 1,500+ managers globally, and at the center of how coaching conversations among peers and between levels get conducted. We'll take a closer look at two IML components, Key Business Concepts and the Employee Development Matrix, to understand how WWT's drive toward a goal of "world class culture" serves as a flywheel that leads to further *innovation for diversity*.

Key Business Concepts

Every manager is expected to understand and master eight concepts that govern how WWT team members are supposed to carry themselves internally and externally, and how they manage their operations. Seven concepts are all predicated on the first: "Core Values." Employees at all levels, especially leaders in the C-suite, hold close THE PATH—Trust, Humility, Embrace Change, Passion, Attitude, Team Player, and Honesty. The first two values, Trust and Humility, are prioritized as the basic tenets upon which the other values are sustained.

In the context of building a culture of innovation and diversity, there are two other Key Business Concepts worth noting: "Always Face Reality" and "The Right People." Teams at WWT are encouraged to confront "brutal facts" of a present situation and to "start with an honest effort to determine the truth." This particular concept is crucial in problem solving and avoiding the type of inertia we discussed in Chapter 3 that is a severe limiter to progress and innovation. Too often when businesses fail to confront difficult truths, problems escalate and become far more costly to solve and allow competitors to exploit weaknesses. The Right People concept is a nod to Jim Collins' *Good to Great* notion that any high-performing organization needs the best people in the right jobs, empowered to make good decisions. More than just having the right people on board, WWT is vigilant about letting go of the wrong people, those who are so toxic and destructive that strong performers leave. This latter concept feeds into another component central to the Integrated Management and Leadership system: the Employee Development Matrix.

MEASURING THE VALUES/PERFORMANCE DUALITY: THE EMPLOYEE DEVELOPMENT MATRIX

Have you ever worked in an environment where company leadership talked a good game of values, but when push came to shove routinely turned a blind eye to toxic behavior and rewarded "performance"? Perhaps it was a sales manager who was poisonous to the culture, but if they exceeded revenue goals, they would be given a pass? It is no surprise that when leaders don't back up value "words" with actions, trust breaks down and any attempt to build a healthy corporate culture unravels. While results in the short term might look good, the longer-term impacts can lead to a company's down-fall. Once the disconnect between values and behaviors takes root, the underpinnings required for innovation—courage, trust, risk-taking, and collaboration—collapse. And great people find opportunities elsewhere.

WWT reconciles the values-performance duality by holding staff members accountable for both, and by evaluating each equally through regular reviews. During orientation, each new team member is introduced to the Employee Development Matrix that plots Core Values along a vertical axis and Job Requirements along the horizontal axis. Job requirements are common across all functions and levels and capture qualities expected of everyone, such as Creativity & Innovation, Quality and Quantity of Work, Communications, and Effectiveness. The Employee Development Matrix not only serves as a construct to evaluate and coach existing employees, but also assesses prospective employees and helps to counsel out employees who fall short on either axis. What makes it work is a track record of managers willing to have difficult coaching conversations just as often on values as on performance. And the fact that individuals who don't perform against the values spectrum will be asked to leave, just as those not performing against business goals.

CEO Jim Kavanaugh will tell you that WWT's success with the Integrated Management and Leadership approach is in part due to the rigor and discipline with which it is implemented across the organization, and the consistency of delivery over two decades. He likes to say that IML is not a "business book of the month club, *it's a long game*." And while there is no question IML is embraced by C-suite leaders at WWT, all 1,500+ business leaders are trained to teach IML concepts to others across the company. So embedded is IML that most sales pitches to new potential customers begins with an overview of the culture that underlies WWT's high level of staff retention, decision-making, well-considered risk-taking, and technical innovation—ideas as important to external clients as to internal staff members.

How the Principles of WWT's Culture Supports Innovation for Diversity

While the contributions of diversity, equity, and inclusion to WWT's business were well understood by both Dave and Jim for years, efforts to build a comprehensive DEI strategy accelerated in 2017 with the hiring of their first chief diversity officer, Bob Ferrell. To signal the importance of DEI across the company, Bob reports directly to Jim and partners closely with WWT's head of global human resources, Ann Marr.

In keeping with one of WWT's eight key business concepts, "Always Face Reality," Bob first set out to understand the baseline diversity mix of employees at all levels, across divisions and regions. He developed a detailed dashboard that monitors weekly changes in employee composition at the business-unit level against goals, as well as the diversity of the applicant pool by job category. As we saw in the overview of the NHL earlier, not as many companies and organizations have the type of sophisticated employee diversity tracking systems one might expect. What is unusual about WWT's system is not only its granularity, but how the information is consumed and acted on at the highest level of the company. Every Monday morning Jim hosts a two-hour executive team meeting where Bob's diversity report is the second item on the agenda. Bob notes that "Jim and the other executives don't let up" on their relentless questioning around attrition, intake for new hires, and areas for improvement. It was out of these weekly executive sessions that new ideas emerged for stepping up progress, such as targeted recruitment strategies, expanded awareness of DEI concepts, and identifying high-potential diverse talent for leadership development. Every C-level leader contributes to DEI problem solving.

Bob, with the support of HR and managers across the organization, aims to improve diversity composition by one percentage point every year. That may sound like modest incrementalism but consider a company of 10,000 employees that is growing new staff positions by 5 percent every year. If it begins with a staff composition that is 10 percent diverse (based on the measure defined by the company), a 1 percent improvement every year will yield nearly 3,000 diverse staff members in 10 years on a base of just over 15,500 employees, or a total percent nearly double that of the starting base. That's progress, especially considering that so many hiring and retention programs focus on improving diversity that *may sound ambitious* lose steam with fading commitment and competing priorities, producing far less impressive outcomes.

Training as Innovation

When Bob recommended organization-wide DEI training on topics as varied as empathy and the consequences of unconscious bias, they didn't hire a group of consultants to conduct training. They did something far more innovative. Leaders from across WWT came together and, with Bob and Ann's support, internally developed a three-hour course that uniquely builds upon IML principles and then sought feedback from across the organization on content. Bob further explained, "We wanted all of WWT's Employee Resource Groups to test drive the material so we could incorporate their input before we finalized. We piloted the training with sample groups across Administration, Sales, and Supply Chain and received honest feedback about what we needed to adjust." Bob and the executive team plotted a three-year plan for training every individual, starting first with all executive leaders then managers. Importantly, as with IML, the course is taught by the executives themselves, as well as a small number of staff members trained in how to teach the material. With such careful intentionality, it's nearly impossible for any employee to dismiss WWT's DEI training as a "one-off" when it is thoughtfully woven into the cultural fabric, spearheaded by top leadership, and with principles incorporated into employee performance reviews.

Radical Listening as an Innovation Pathway

Another cornerstone to WWT's DEI and culture-building construct— and a source of inspiration for further innovation for diversity—is the series of employee listening sessions that began after the murder of George Floyd at the urging of a group of Black employees. When we first heard the term "listening sessions," our first reaction was, "sure, a lot of companies did this." But like so many other efforts to carefully scale culture at WWT, their listening sessions were designed with an action-orientation and the full attention of executive leaders.

Bob shared that from June 2020 through September 2022 over 260 90-minute listening sessions took place with 15–25 employees in each. In the early days of these sessions, the impacts of racism, justice, and community reactions were central themes. When the pandemic hit, discussion topics turned to mental health and wellness, impacts on school-aged children, and remote working. Hate crimes against Asians was a focal point during several sessions. Bob and his team regularly survey employees on issues they want to talk about, and topics get published in advance so anyone can join.

What makes these listening sessions different from those in many other companies is the level of executive engagement, commitment to action, and transparency. Both a sponsoring senior executive (including the CEO, COO, and CFO) and a member of the DEI attend *all of them.* The role of the senior exec is to set up the conversation, then step back and listen. Bob notes that the role of the DEI team member is to, "summarize what we heard, identify themes across sessions, surface the 'big rocks' we have to tackle, and finally work with an 'action owner' on a plan." During Jim's Monday morning executive team meetings, each senior leader reports on what they heard from the prior week's listening sessions and the issues that need to be addressed. Every quarter the entire WWT employee base is informed of concrete actions underway to address repeated themes, and progress against those actions.

Ann Marr noted that the executive accountability for engaging and acting on these listening sessions further deepens employee trust, openness, and communication between levels and across functions. The benefit is that when complications arise in day-to-day operations, there is a higher likelihood that team members will surface problems more quickly and be more engaged in problem solving.

Creating Career Pathways for All Employees

Forbes reported that through early 2022 the average turnover for tech professionals was over 23 percent; in other words, nearly a quarter of a company's technology employees must be replaced in one year. According to a 2021 study by Gallup, voluntary turnover costs U.S. businesses over $1 trillion each year, but the monetary damage is only the beginning. Rewiring the personal connections across an organization when new employees are hired, not to mention onboarding and training, take a heavy toll, especially at any company focused on improving culture and operating at high velocity.

WWT prides itself on a voluntary turnover rate among its professional and technical teams of well under 10 percent—almost unheard of. While the culture and values—the IML system—are certainly significant factors in the high retention levels, well-articulated pathways for career advancement are another motivation to stay and grow with the company. In 2020, Ann Marr and her HR team launched a comprehensive "Career Development Framework" that codifies the skills and competencies for virtually every role in the organization. Managers are expected to routinely discuss career aspirations with team members and plot short- and long-term goals based on an employee's present skills, and those required to qualify for lateral or more senior positions. WWT makes a series of self-directed training opportunities available for motivated employees to upskill and qualify for new opportunities with higher pay. Since every job opening posted details the skills and competencies as determined by the Framework, any employee can see how their present skills match up, as well as the additional course work or experience needed to be a qualified applicant.

Ann noted how the Framework facilitates practical conversations with team members eager to take on new opportunities, "The onus is really on the employee to signal to their manager what they want to do in their careers. We want people to be invested in their own growth and empowered with access to the tools, training, and support to reach their long-term goals."

Career Pathways as a strategy to increase retention of hourly staff. The transparency for how to move up and across the organization has been highly motivating for WWT's professional and technical teams, and certainly contributed to retention. However, in other parts of the organization, turnover was more problematic. Almost a third of WWT's employee base is comprised of hourly shift workers in warehousing and logistics roles where job tenure is approximately 25 percent less than the average tenure of salaried employees in technical and administrative roles.

As part of a strategy to increase retention for hourly team members, WWT rolled out the same set of tools and trainings to those in warehouse and logistics roles. A warehouse associate level 1, for example, knows exactly what it takes to advance to levels 2 and 3. Or, to cross into IT roles within the WWT Advanced Technology Lab, which is an adjacent career pathway, and then potentially into more skilled roles within WWT's professional services division. Warehouse and logistics associates are given four hours every week to pursue training to receive the additional skills and certifications required to advance. And, as with the technical staff, managers are encouraged to discuss with each warehouse and logistics staff member their career aspirations.

Since the composition of the hourly workforce at WWT is more diverse among a number of dimensions, their "recruit technical talent from within the ranks" strategy also promotes an influx of different perspectives and backgrounds.

Within just three months of launching the Career Development Framework program with the warehouse and logistics teams in June 2022, over 10 percent of associates had already voluntarily engaged with the training content. During those first three months, associates logged a total of nearly 3,000 hours of online coursework, with IT certifications, programming languages, spreadsheet development, and personal growth (including communication skills) among the most popular classes.

Ann emphasized that providing opportunities for hourly workers to advance and grow at WWT is only one approach to increasing retention among this employee base. She reflected on how not all employees are motivated by the same priorities, "Not everyone wants to climb the ranks, and that's okay. We need to listen to the needs of every part of our workforce and find ways to make their experience at WWT better. We can't always accommodate everything, but we pay attention to the big themes." One example she cited was the recent investment in on-site medical clinics at each of WWT's warehouse sites to support preventative and routine healthcare for those working shifts that made keeping medical appointments challenging. Access to free prescription medications and a regular doctor was not only a convenience, but it also provided life-saving interventions for numerous employees who otherwise would not have gotten serious medical conditions diagnosed.

The Career Development Framework implemented throughout the entire company is an important tool to reinforce and build on one of WWT's "key business concepts" we discussed earlier: The Right People. Why wouldn't a company want to invest in and advance talent so carefully hired for culture?

Any one principle of WWT's keystone to its culture, Integrated Leadership and Management, would be considered routine at any successful company. The fact that all eight principles, including core values, are now encoded into the company's DNA through continual reinforcement, executive action, training, performance reviews, and decision-making is what allows WWT's unique culture to scale. The IML system advances the core principles we introduced in Chapter 3—courage, collaboration, trust, risk-taking, and especially leadership—necessary for a culture that encourages innovation. Including innovating for diversity.

In our discussion of both the NHL and WWT we examined comprehensive, systems-level commitments to embedding DEI values across their organizations. The scope of these commitments are driven by the courage and actions of C-suite leaders. In Table 10.1 we review how the actions summarized in these two case studies reflect our principles of innovation and avoid innovation threats.

Table 10.1: Evaluating Systems-Level Change Strategies against Innovation Principles

PRINCIPLE NECESSARY FOR A CULTURE OF INNOVATION	PRINCIPLE IN ACTION
Courage	▪ The leadership at the NHL had the courage to confront the broader, more substantial impacts on the overall business of the League if they did not invest in a comprehensive DEI strategy.
	▪ WWT demonstrated courage when establishing the Employee Development Matrix that requires managers to make the difficult decision to exit an employee who may be performing against job requirements, but disregarding core values.

Continues

Table 10.1 (*continued*)

PRINCIPLE NECESSARY FOR A CULTURE OF INNOVATION	PRINCIPLE IN ACTION
Risk-Taking	▪ The NHL took a measured risk when choosing to build relationships with new, non-traditional talent sources, such as Jopwell. ▪ WWT risked vulnerability and the expectation for action when they launched their ambitious small-group employee listening sessions with a high level of executive engagement.
Trust	▪ The trusting partnership built between the NHL Commissioner and the executive leader charged with leading "diversity as a growth mindset" was central to the successful launch and ongoing progress of several connected DEI initiatives. ▪ The two WWT founders have modeled from the inception of the company the power and the expectation of earning and growing trust among colleagues and with clients.
Collaboration	▪ The understanding at the NHL that a wide-ranging diversity strategy required collaboration with both the teams and their owners, and with the Coaches' Association. ▪ Recognizing they could not build a youth hockey movement alone, the NHL developed a collaborative partnership with an expert, RCX Sports.
Leadership	▪ At both the NHL and WWT, leaders at the most senior levels personally committed to systems-level change required to "scale" culture, and to quantitatively track progress against goals and benchmarks.

INNOVATION THREATS	ACTIONS TO AVOID THREAT
Low Priority	Concerns connected to growth mobilized full-scale drives toward inclusion at both the NHL and WWT. For the NHL, building a new, diverse and energized fan base was key to igniting long-term growth of professional hockey. For WWT, their rapid business growth could have compromised their commitment to a culture of diversity, but their intentional prioritization of DEI instead served to scale that culture as their staff numbers skyrocketed.
Inertia	The NHL is building on the momentum of their successes to propagate a "Movement, Not a Moment" message that embeds DEI thinking and actions at every level of their organization.

Leaders at WWT might have been satisfied with diversity gains and low turnover among their technical/professional teams, but instead chose to invest resources within their warehouse and logistics division to accelerate career growth and retention among warehouse associates who represent greater diversity across multiple dimensions. |
| Arrogance | NHL executives could have been satisfied with solid sponsorship and fan numbers in recent years, but instead committed to a long-term plan across seven dimensions (Leadership, Education, Employment, Marketing, Partnerships, Community & Civic Engagement, and Youth Participation), all of which have embedded a diversity orientation, geared to sustainable growth.

In spite of WWT's exceptional business success, it remains deeply committed to principles of fairness, equity, and performance forged early in the company's founding and codified in its Integrated Management and Leadership system, which is required training for all supervisors. |

Conclusion

In this chapter we examined how the public and internal expression of commitment to DEI among CEOs and their executive teams is necessary but not entirely sufficient for diversity practices to be successfully embedded within an organization. Leaders who adopt a strong promotion focus and resource-based view of diversity over the long term have a higher likelihood of operationalizing practices at all levels that, over time, make DEI a central part of the culture. We reviewed the examples of the National Hockey League and World Wide Technology as two organizations where leaders, starting with the CEO, are taking a comprehensive, long-term approach to building systems that bolster inclusive cultures and build greater value for all stakeholders, especially employees. Leaders at both organizations acknowledged that their systems weren't perfect and spoke of being on a journey, learning every day how to improve.

Summary

- While the words and actions of a CEO are vital for DEI practices to succeed in any organization, more important is a long-term commitment to the systems that operationalize those practices across operations, as well as an intrinsic belief that diversity yields business value.

- Effective CEOs hold managers accountable for DEI practices and outcomes just as they would any core business priority.

- Leaders at both WWT and the NHL empowered executives with building the structures and processes to embed diversity principles within the context of their respective cultures, rather that outsourcing that work to external parties. As such, in both cases there exists a high degree of ownership. "Diversity" is not a stand-alone program or initiative.

- Steady, consistent progress against diversity goals that may appear incremental can be far more effective in the long term than showy, ambitious goals that lack persistence. Leaders at both the NHL and WWT spoke of the importance of playing a long game with focus and intention.

- As demonstrated by the NHL's *seven dimensions of excellence* and WWT's *integrated leadership and management* system, organizations with strong cultures that prioritize diversity and innovation codify behaviors and expectations.

- The core principles introduced in Chapter 3—trust, humility, courage, collaboration, and especially leadership—are central to both the DEI journeys of WWT and the NHL and to their cultures.

Final Thoughts and Next Steps

Throughout the book, we have presented firsthand accounts of how leaders across industries, functions, and organizational size have applied innovation principles to launch and accelerate their DEI efforts, building strong, inclusive environments where *all* employees thrive.

While we hope these stories are inspiring, you may wonder how you can take the learnings from the case studies and apply them in your own organization.

In this final chapter, we will outline the key takeaways from our interviews to encourage a series of provocative conversations in your organization and offer guidance for applying innovation principles to your own real-world DEI practices.

Key Concept: Innovating for diversity does not follow an established formula. Taking the insights, inspiration, and lessons learned from the previous chapters we offer a set of questions for you to explore as a starting point to your own DEI innovation journey.

Lessons in Innovating for Diversity

In our research and through our conversations with leaders at several forward-thinking organizations, we discovered several common themes across their approaches for diversity innovation. Chief among them are the following:

- **CEO and C-level support is not enough**

  Visible CEO and C-level support is necessary but not sufficient for making diversity an actionable priority. Public proclamations that announce ambitious goals for a more diverse and inclusive workplace can galvanize other leaders across industry sectors to be similarly bold. They can also ignite individual leaders within an organization to propose new ideas. However, such proclamations require clear plans that are operationalized at every level for success to take root.

- **DEI must be an integral part of organizational culture**

  DEI must be an integral part of organizational culture where supporting behaviors and expectations are codified. Like with any set of values that describe the desired culture of a company, the actions, and behaviors of individual leaders to advance diversity within their teams must be visible and evaluated. Reinforcing behaviors should be celebrated and harmful actions addressed with clear feedback and course corrections. Organizations with innovative DEI performance records are skilled at embedding practices (often internally developed) that are consistent with their cultures and characterized as business imperatives. They are also expert at identifying individuals up and down the organizational hierarchy who are "culture carriers" and who lead by example. When considered integral to a company's culture, DEI also then becomes a part of the company's DNA in a way that survives personnel changes or shifts in day-to-day business priorities.

- **Measurement matters**

  Important key performance indicators and metrics should be identified at the start of implementing any program or initiative. Establishing baseline quantitative and qualitative data, together with regular progress checks, allows organizations to quickly identify areas for further review and improvement. As we saw in several of the case studies, baseline employee or recruitment pipeline data often doesn't exist. But, once an organization has done the heavy lifting to establish these baselines, developing goals and measuring progress is far more straightforward. As we saw in Chapter 9, DEI is not prioritized unless it is vigilantly monitored and measured like any other business deliverable, such as sales figures, market share, or brand awareness. If dimensions of DEI are not vigorously measured, they do not matter.

■ **Accountability is required**

Once measures and goals are established by leaders, organizations certainly should celebrate success. Acknowledging success builds momentum, and shapes what excellence looks like. When failures occur—which will inevitably happen—it is essential to take the time to examine root causes and evaluate necessary changes. Without a consistent focus on accountability, coupled with training and the appropriate resources, overarching DEI efforts can regress in importance to other goals perceived as more urgent. Acknowledgment of failures and struggles, and taking responsibility in improving outcomes, is a sign of a healthy culture and is a requirement for DEI gains.

■ **DEI work must be collaborative across an organization**

DEI work isn't the domain of just one leader or one department. As we have seen across several of our case studies, companies successful in their DEI practices encourage collaboration across several departments or business units, and among those of varying levels of responsibility and tenure. They are also humble enough to acknowledge where they lack knowledge and expertise. Rather than create DEI programs in a siloed fashion, they engage with subject matter experts and their employees (and in some cases vendors and customers) to thoroughly understand root problems and innovate to create better outcomes. To place the burden entirely on the chief human resources officer or chief diversity officer is akin to expecting innovation to be the sole responsibility of a chief innovation officer, and discouraging employees working in all corners of a company from contributing ideas for continuous improvement.

■ **It's okay to start small**

Pilots and experimentation are central to any innovation. Many of the companies we profiled in Chapters 4 – 9 took the time to carefully construct controlled pilot programs to test new approaches to improving their DEI outcomes, whether around recruitment, mentoring, apprenticeships, or retention. Starting small afforded several advantages: opportunities to learn what works well and what needs to be improved, lower initial costs, and the time to socialize and build support for the test with others across the company. If successful, a pilot program can spark other new ideas and improvements and elicit discussion on how to scale the program to other parts of an organization. As we saw in the case studies on Citi's apprenticeship program (Chapter 4) and Northrop Grumman's skill-based hiring effort (Chapter 6), the pilots produced a "trickle-up" effect that led to changes in practices and policies in other areas across the companies and encouragement for further innovation to advance diversity.

▪ **Change underlying systems that block progress**

Know when and how to tackle underlying systems limiting DEI progress. Starting small and systemically tackling components of HR or business practices through innovative pilots can often reveal the Fixed Practices and Fixed Attitudes (Chapter 1) that are entrenched to the point that they are nearly invisible. Only when those fixed qualities are addressed by stakeholders as root problems can DEI programs scale and thrive. In the case of Tessco, we saw how intentional efforts to hire, onboard, and develop diverse entry-level talent provided a strong pipeline for director-level roles. Their former practice of hiring mostly experienced, mid-level talent reduced the size and diversity of their applicant pool but was such an embedded practice that it went unquestioned until a new leader with a fresh perspective came on board.

While we admire the ambition of headline-grabbing announcements many CEOs have made in recent years, they become empty promises without changing the underlying systems that produced the results in need of improvement.

▪ **Steadfast Commitment is key**

The social unrest resulting from George Floyd's murder and hate crimes against the Asian American Pacific Islander (AAPI) community provided the impetus for many corporate leaders to reexamine their approaches to DEI. Yet, all organizations must understand that a commitment to DEI cannot be tied to moments in time alone. To be clear, we should never forget the horror and atrocity of these moments. Ever. But for DEI programs to be meaningful and sustainable, they must transcend specific events and be anchored in both clearly understood goals and woven into culture. Senior leaders and their teams must commit to DEI *every day and over the long-term*. As we learned from JT Saunders of Korn Ferry and Kim Davis of the NHL in Chapter 9, CEOs with a resource-based view of diversity lead with courage and tenacity, and do not let up when the business cycle takes a downturn.

▪ **Strong leadership matters**

Successful DEI practices fueled by innovation focus on the individual, understand employees' unique skills and interests, create a culture and environment that supports success, leverages passionate and determined leaders who are energized by unlocking potential in others. Leaders comfortable with these four ideals are more likely to invest personal capital—time, reputation, connections—to champion novel approaches to improving diversity and inclusion, and to challenge Fixed Attitudes and Fixed Practices.

▪ **Provide tangible support and resources**

Diversity and inclusion efforts, without attention to equity, may ring hollow. To retain and advance diverse talent, they will need tangible tools, resources, and support from leaders in their career journey. While this support can include mentoring, strong sponsorship, and autonomy over how they perform their work, a comprehensive approach to career development and advancement for diverse talent is key.

▪ **Have clear understanding of core problems**

DEI is an imperfect journey, not a destination or checklist. Even with their accomplishments, the organizations we spoke with acknowledge that their efforts have not always been executed flawlessly, nor that their level of success is a reason to rest on their laurels. They continue to learn from their progress and their setbacks and apply the principles of innovation to make continuous improvements. Leaders at these companies also realize that effective solutions rarely come in the form of a DEI checklist but require careful analysis and creativity to arrive at an answer that best fits their needs and culture.

Just as few successful businesses adopt a simple checklist approach to improving sales or scaling operations, those who achieve strong DEI outcomes have a clear understanding of the core problem they seek to address and tailor their solutions accordingly. Furthermore, understanding the relentless focus DEI requires prevents the threats to innovation we discussed in Chapter 3, Low Prioritization, Inertia, and Arrogance, from hindering the emergence of innovative ideas and sustained progress.

Revisiting the Virtuous Cycle of Innovation and Diversity

The word "innovation" may evoke purely technological or scientific concepts, or suggest an idea primarily connected to technical products and services. However, the definition of innovation we subscribe to, taken from the government of New Zealand, suggests a more nuanced meaning: "Innovation is the creation, development, and implementation of a new product, process, or service, with the aim of improving efficiency, effectiveness, or competitive advantage."[1]

As we saw in Chapter 3, substantial research exists on the undisputed link between diversity and innovation: more diverse teams with a greater range of cognitive approaches, lived experiences, expertise, and backgrounds lead to greater success in innovating new processes, or services. We have presented examples throughout this book of how leaders have in turn applied principles of innovation to successfully build more diverse teams. For the leaders we spoke with, they examined the challenges that have historically plagued DEI programs and reframed their approaches; they utilized principles of innovation—Courage, Trust, Risk-Taking, Collaboration, and Leadership—to achieve success through novel approaches. This notion of a reinforcing interplay between innovation and diversity is represented in Figure 10.1.

Some of the case studies we presented emphasized one innovation principle over others and some touched on the five principles altogether. This demonstrates that not all needs to necessarily exist, but the likelihood of success in DEI efforts increases when more of the principles are present. For example, having a lack of the Leadership principle alone makes it

---

[1]New Zealand Ministry of Business, Innovation and Employment. (2019). (rep.). Deepening our understanding of business innovation (pp. 10).

highly unlikely that a new DEI approach will thrive. Lacking the Trust principle can also make it unlikely that leaders, DEI practitioners, and employees feel safe in expressing critical feedback or vulnerability. And without the Risk-Taking principle, we are likely to continue to invest in one-off compliance-driven DEI programs that will not produce sustainable results or underlying systems change.

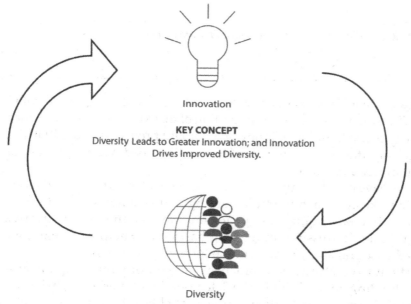

KEY CONCEPT
Diversity Leads to Greater Innovation; and Innovation
Drives Improved Diversity.

Figure 10.1: The virtuous cycle of diversity

As important as the five core principles are to innovation, we believe leaders—those who formally hold the title, who serve in the role informally in their departments, and emerging leaders—must equally focus on preventing the three innovation inhibitors from undermining their progress: Low Prioritization, Inertia, and Arrogance. A consistent theme that emerged from our interviews reinforces how easy it is for corporate leaders to become distracted by short-term priorities or to become satisfied with modest gains and lose the resolve to attain more ambitious goals and permanently weave DEI practices throughout their companies. Those leaders who view DEI with a growth mindset and value its contributions to overall business outcomes will continue to prioritize efforts, overcome inertia, and apply humility to the process of continuous improvement. As we saw with WWT's success in scaling culture with diversity at its core, leaders must be unwavering in their commitment over the long haul.

Applying Innovation Principles in Your Organization

"Knowledge is of no value unless you put it into practice."

~ Anton Chekhov

Now it is your turn: We are firm believers that the sharing and distribution of knowledge is good; the application of that knowledge is *even* better.

It may seem daunting to innovate your DEI programs and initiatives, but getting started on the path of innovation need not be difficult, or prohibitively expensive.

As we have noted throughout the previous chapters, we do not advocate a recipe-like approach to implementing and sustaining initiatives to support diversity, equity, and inclusion. Innovation requires creative thought and addressing underlying barriers. To help you begin, we instead encourage you to answer the following questions:

1. What fixed attitudes or practices are you or your company currently holding on to? Thinking back to your own experiences, have you or people in your organization demonstrated Fixed Practices? Some examples are reflected in phrases like:

- "Change takes a long time."
- "We've tried a diversity program before."
- "We already have a Chief Diversity Officer."
- "Changing HR recruiting systems is expensive."
- "We really need to recruit for a narrow set of specialized skills."
- "Our business is doing fine, why risk making any changes?"
- "Our HR department is handling unconscious bias training and runs some ERGs—isn't that enough?"
- "I understand there is a 'business case' for diversity, but right now I'm more worried about inflation."

Additionally, have you experienced or observed the signs of Fixed Attitudes? Can you think of times when you or others expressed fear, indifference, denial, or anger? What were the contributing or triggering factors that lead to these emotions? How can you mitigate or challenge the influence of Fixed Practices and Fixed Attitudes and replace them with new mental models?

2. Where have you seen innovation applied in any part of your organization? What does your company hold up as one of its most successful innovations? What roles did Courage, Risk Taking, Trust, Collaboration, and Leadership play in activating that innovation through to implementation? Consider how you can harness the momentum from that innovation to a brand-new approach connected to a DEI-related opportunity. Further, how might the individuals across the organization best known for innovation and creative thinking be engaged in "innovating for diversity?" Encouraging employees from across the company to offer fresh perspectives, borrowing from their own firsthand experiences, can deliver insights that complement more traditional DEI practices.

3. What opportunities exist in your organization to apply any of the specific tactics described in the prior chapters? Hopefully, the stories and tangible examples you have read here have provided a bit of inspiration. While we are advocates of creating tailored solutions to meet the unique DEI challenges specific to any one organization, you may find that there are new paths to explore based on the case studies we shared. If finding talent with precisely the right skills is your challenge, consider launching an apprenticeship program customized to your needs. If you believe in the importance of mentorship but your programs are flailing, can you apply the innovations learned from the team at Zendesk? Are you struggling to retain diverse talent? Consider how you might best evaluate root causes of turnover. Can you create new approaches to career development, workplace flexibility, and benefits offerings, if those dimensions turn up lacking, as they do for so many companies as we learned in Chapter 8.

4. What are some strategies and tactics you can use to minimize innovation threats? Inertia, making innovation a low priority, and arrogance can derail even the most successful DEI programs. As we, at the time of this writing, are entering into a period of economic uncertainty and inflation, it can certainly be tempting for companies to redirect their energies and financial resources elsewhere.

Take time to think through ways to turn these potential threats into opportunities. Are there allies within the organization that you can lean on for support and advocacy in your efforts? Are there studies or external resources you can leverage to illustrate the *opportunity cost* of inaction? We know from our findings in Chapter 3 that smart companies that persist with innovation investment throughout economic downturns are better positioned to outperform their competitors during recessionary periods and emerge stronger in the early years of economic recovery. We believe the same potential exists for companies that continue to *innovate for diversity* during downturns. These are the companies that will succeed in building culture, loyalty, performance, and world-class teams prepared for growth once markets open.

We recommend brainstorming answers to these questions with your respective teams and leadership. Use these discussions to establish a tone reflective of the innovation principles and as an opportunity to connect as a team on how together you can uncover and address the Fixed Attitudes and Fixed Practices that may prevent you from achieving success. And remember, you do not need to solve every problem all at once. Pick the area where innovation will bring the greatest impact, create baselines, be creative, establish accountability, pilot, learn, and adjust. And be fearless. We encourage you to come to our Linkedin group to share and further the discussion at `www.linkedin.com/groups/12752095`.

We wish you all the best as you dive in and innovate for diversity.

Index